Structural Engineering Failures:
lessons for design

By

Niall F. MacAlevey

Dedication: This book is dedicated to my parents.

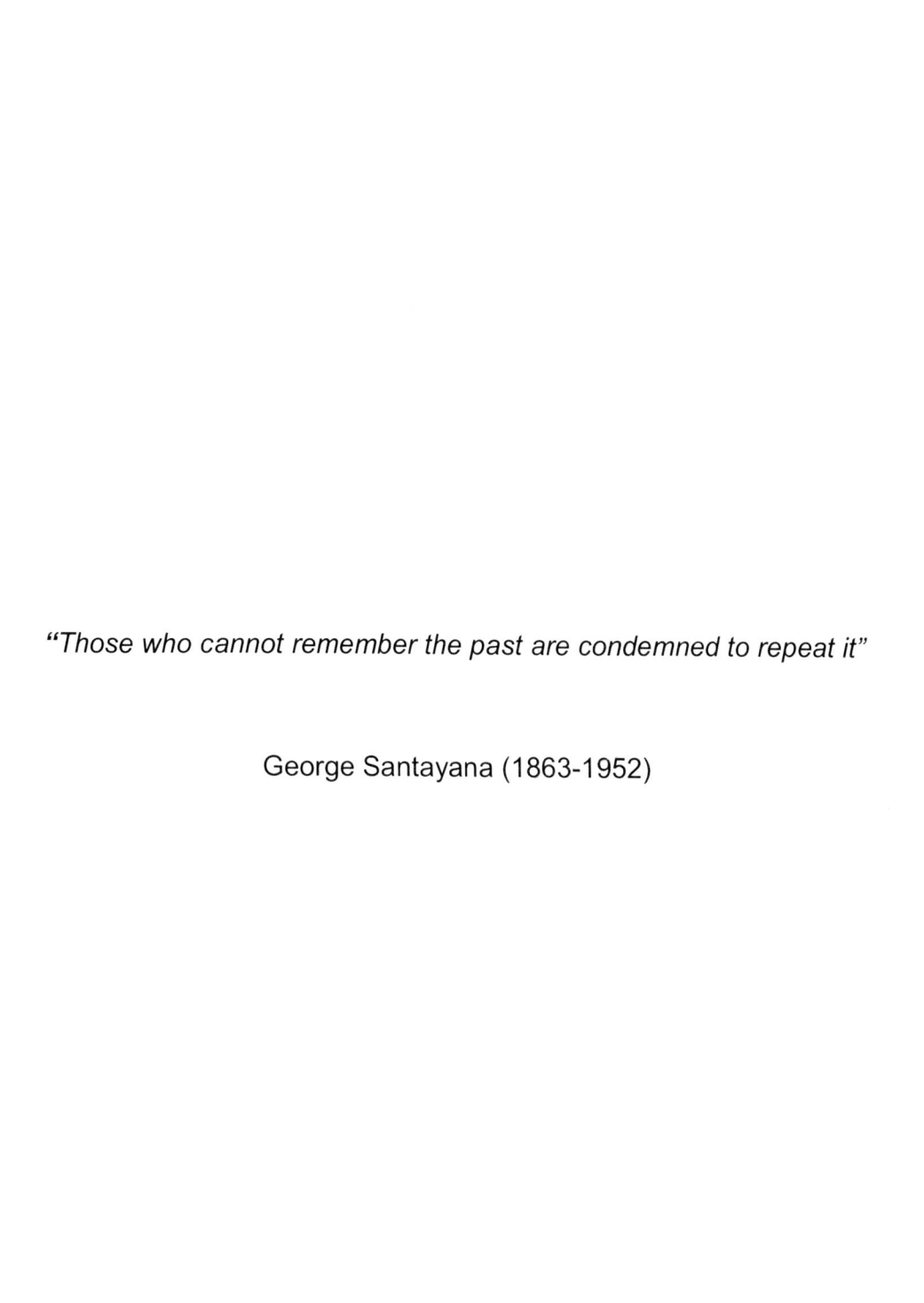

"Those who cannot remember the past are condemned to repeat it"

George Santayana (1863-1952)

Preface:

Collapse is every structural engineer's biggest fear. Fortunately not every engineer must go through the experience of a collapse first-hand: we can all learn from the mistakes of others. In fact, it has been said that the *only* way in which we can prevent structural engineering failures is by more education (Kaminetzky 1991). This is the purpose of this book.

We can learn least from structural engineering successes, as these structures are not tested. Structural failures, on the other hand, are full-scale tests. If something goes wrong it is essential that the failures are studied so that the true circumstances of the failure (and thus the nature of the test) established so that we can learn what *not* to do next time.

Many case studies of the failure of Civil Engineering Structures throughout the world are presented. These case studies are grouped into several chapters each containing case studies on a particular topic. The divisions are arbitrary and purely for organizational reasons. Lessons are drawn from each case study. Only non-earthquake related failures are considered.

By preparing this book I hope to encourage engineers to do the following:
- Design structures that are more resistant to collapse.
- Avoid making the same mistakes other engineers made in the past.
- See the importance and limitations of checking and of codes of practice.
- Be aware of danger signals.
- See how preventable failures are.
- Learn more about engineering judgement.
- Encourage a more reflective approach among designers so they don't just apply the code blindly.

- See the relationship between analysis and design.

As will become evident I hope engineers will look at design more critically and ask questions like:

Is the minimum factor of safety recommended by the code appropriate for this structure (or part of it)?

What is the likely collapse behavior of the structure: ductile or brittle? How can it be improved?

The contents are as follows: Page

1. Introduction   9
   1.0   Failures.   9
   1.1   Progressive Collapse.   11
   1.2   Ductile and Brittle.   13
   1.3   Statical determinacy.   14
   1.4   Code approach.   17
   1.5   Tied solution.   19

2. Gravity
   2.1   Walkways at Hyatt Regency Hotel, Kansas, USA, 1981.   21
   2.2   Hotel New World, Singapore, 1986.   25
   2.3   Sampoong Department Store, Seoul, Korea, 1995.   28
   2.4   Terminal 2, Paris Airport (Charles de Gaulle), France, 2004.   33

3. Computers & Modelling
   3.1   Hartford Civic Centre, Connecticut, USA, 1978.   37
   3.2   Sleipner A offshore platform, Norway, 1991.   42
   3.3   Ramsgate Walkway, Port of Ramsgate, UK, 1994.   47
   3.4   Compassvale School, Singapore, 1999.   51

4. Wind
   4.1   Tay Bridge, Scotland, 1879.   57
   4.2   Tacoma Narrows Bridge, Washington, USA, 1940.   60
   4.3   Ferrybridge Cooling Towers, Pontefract, UK, 1965.   65
   4.4   John Hancock Tower, Boston, USA, 1972.   69
   4.5   Citicorp Centre, New York, USA, 1978.   75

5. Maintenance
   5.0   Introduction   80

15	5.1	Point Pleasant Bridge, Ohio/West Virginia, USA, 1967.	84
16	5.2	Mianus River Bridge, Connecticut, USA, 1983.	87
17	5.3	Antioch School, California, USA, 1980.	92
18	5.4	Ynys-y-Gwas Bridge, West Glamorgan, Wales, UK, 1985.	94
19	5.5	Schoharie Creek River Bridge, New York, USA, 1987.	97
20	5.6	Piper's Row car park, Wolverhampton, UK, 1997.	100
21	5.7	De La Concorde Overpass, Montreal, Canada, 2007.	103

6. Welding

22	6.0	Introduction.	110
23	6.1	King's Street Bridge, Melbourne, Australia, 1962.	115
24	6.2	Brooklyn College, New York, USA, 1971.	119
25	6.3	Alexander Kielland, Norway, 1980.	122
26	6.4	SMRF, Northridge, California, USA, 1994.	125

7. Construction

27	7.1	Quebec Bridge, Quebec, Canada, 1907.	129
28	7.2	West Gate Bridge, Melbourne, Australia, 1970.	133
29	7.3	Bailey's Crossroads Condominium, Virginia, USA, 1973.	139
30	7.4	Willow Island Cooling Tower, West Virginia, USA, 1978.	143
31	7.5	Cocoa Beach Condominium, Florida, USA, 1981.	146
32	7.6	L'Ambiance Plaza condominium, Connecticut, USA, 1987.	148
33	7.7	MRT temporary works, Singapore, 2004.	151

8. Fire, Explosion, and Impact

34	8.1	Ronan Point, London, UK, 1968.	157
35	8.2	Sunshine Skyway Bridge, Florida, USA, 1980.	160
36	8.3	Abbeystead Valve House, Lancashire, UK, 1984.	162
37	8.4	Alfred Murrah Building, Oklahoma, USA, 1995.	165
38	8.5	World Trade Center 1&2, New York, USA, 2001.	168
39	8.6	The Pentagon, Arlington, USA, 2001.	175
40	8.7	World Trade Centre 7, New York, USA, 2001.	180

9. Cladding
    9.1 Amoco Tower, Chicago, USA, 1989. — 184
    9.2 HKH A apartment blocks, Hong Kong, China, 1992. — 187
    9.3 Condominium Building, Singapore, 1993. — 188
    9.4 "Big-dig" tunnel, Boston, USA, 2006. — 192

10. Miscellaneous
    10.1 US air force warehouses, Ohio/Georgia, USA, 1955/6. — 195
    10.2 Stepney School, London, UK, 1974. — 199
    10.3 Kemper Memorial Arena, Kansas, USA, 1979. — 202
    10.4 Shopping Centre, Burnaby, British Columbia, Canada, 1988. — 209
    10.5 Koror-Babeldaup Bridge, Palau, 1996. — 212
    10.6 Millennium Bridge, London, UK, 2000. — 216
    10.7 I-35W Bridge, Minneapolis, USA, 2007. — 219
    10.8 D. L. Lawrence Convention Centre, Pittsburgh, USA, 2007. — 222

11. Conclusions — 227

Worked Examples — 242
References — 272
Appendix 1 — 280
Appendix 2 — 282
Index of cases — 285

Chapter 1: Introduction

"Engineers should be slightly paranoiac during the design stage. They should consider and imagine that the impossible could happen. They should not be complacent and secure in the mere realization that if all the requirements of the design handbooks and manuals have been satisfied, the structure will be safe and sound."

Lev Zetlin (founder of Thornton-Tomasetti Engineers).

1.0 Failures

What is the reason for failures?

According to Feld & Carper (authors of "Construction Failures", Feld & Carper 1997) they are all due to either "Ignorance, carelessness, or greed". They note that **all** failures are due to "**human error**". We may speak of "**material failure**" but what we really mean is a failure to design or maintain properly or detect a flaw in its manufacture. These are clearly human errors. A material always behaves according to the laws of physics.

When do most failures occur?

"Total collapses occur most frequently during construction." according to D. Kaminetzky, author of "Design and Construction Failures", (Kaminetzky 1991). He also notes that *"Most structural failures are due to gross errors or omissions. The most important factor which would reduce the amount of failures is **education**".*

W. Ransom, author of "Building Failures", (Ransom 1981) made this observation: *"Knowledge seems to become mislaid from time to time. Those with long memories...are often struck by the re-emergence of problems which have been well researched and documented".*

Referring to the apparent 30-year cycle between major failures Silby & Walker (1977) note that there is a *"...communication gap between one generation of engineer and the next."* (Major bridge failures have taken place in 1847 (Dee), 1879 (Tay), 1907 (Quebec), 1940 (Tacoma Narrows), and 1970 (West Gate/Milford Haven).

Let us note two figures to put the risk of structural failure into better perspective:

- Probability of fatality due to structural collapse: 1/10,000,000 per year;
- Probability of fatality due to vehicle accident: 1/5,000 per year.

So construction is a low **risk** activity? Actually, it's not. The definition of risk is really Probability x Consequences. We are dealing with low failure rates (i.e., the probability low), but the consequences can be high. This means that the risk is not necessarily low.

Duffey and Saull (2003) studied failures in a wide range of industries, from civil engineering to mining to aviation, etc. For each particular industry they found that a similar curve shape described the failure rate versus time. This is shown in Figure 1.1 below.

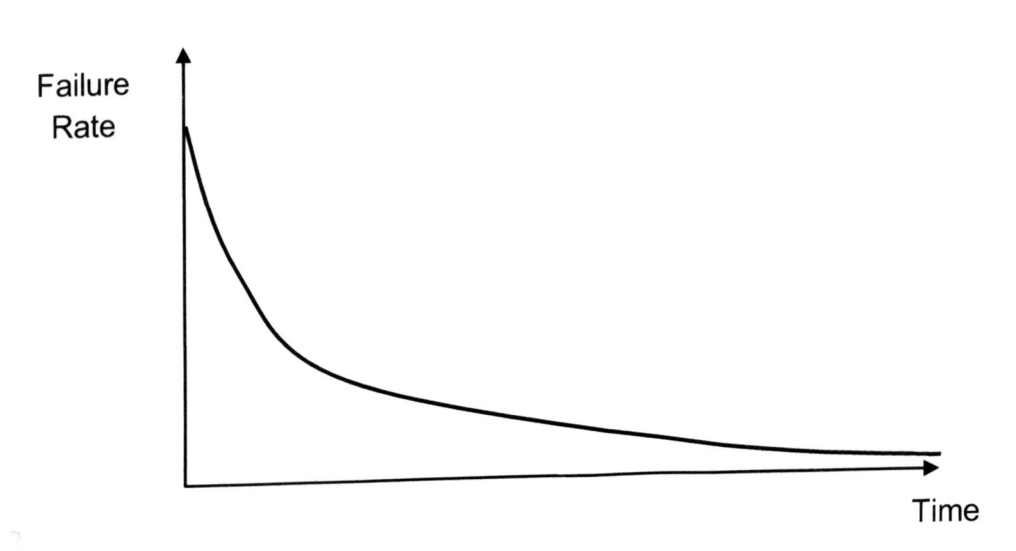

Figure 1.1: Failure Rates over time for a variety of industries

We can see two interesting things from this curve:

1. Failure rates drop over time.
2. Failure rates do not drop to zero.

Railway bridges give a good example of the first item. In the nineteenth century when cast-iron was common failures rates were very high. For example, in the 1880s about 25 railway bridges collapsed each year in USA/Canada. Nowadays failure of such structures is very uncommon. Thus this tells us that we learn from our mistakes. The second observation is less optimistic. There is always a residual failure.

Since the rate is never zero and structures are increasing in size over the years (e.g., sports stadia, high-rise buildings) and we have classes of structures that didn't even exist in the past like shopping malls. The consequences of failure are only likely to increase in the future. Thus we have to be vigilant. According to Charles Perrow *"the threat of industrial and technological disasters is increasing because of increasing concentration"*, Perrow (2007).

1.1 Progressive Collapse

The first thing that we, as engineers, must do is ensure that the structure is *safe*. As far as possible, any defects in the structure must not lead to wide-spread collapse, i.e., we want to ensure that the structure has good resistance to *progressive collapse*. A structure has good resistance to progressive collapse if it can prevent a *local* collapse becoming a *global* collapse. This resistance is also known as "structural integrity" or "robustness" or (perhaps confusingly) "redundancy".

A statically determinate structure has poor progressive collapse resistance. By definition "statically determinate" implies the structure is only one step away from being unstable. Consider a simply supported beam, spanning from A to B and supporting a vertical load.

Suppose vertical support at support B is lost for some reason. Then, no matter how strong the beam, or support A is, the structure is **unstable** and so will collapse.

There is a simple rule to identify these structures: recall that there are just enough members, constraints between members, and supports, to preserve the shape and keep the structure stable and at rest under load. So the rule is:

- If any one member, constraint, or support is removed, then the structure collapses, (i.e., undergoes a *major* change of shape.) then it was statically determinate before the change was made.

However it is not enough to simply ensure that the structure is statically indeterminate, i.e., redundant (this is the meaning of the word "redundant"); adequate **strength** must be provided to ensure the structure is stable in its new configuration.

Consider now a statically indeterminate structure; say a propped cantilever beam (Figure 1.2):

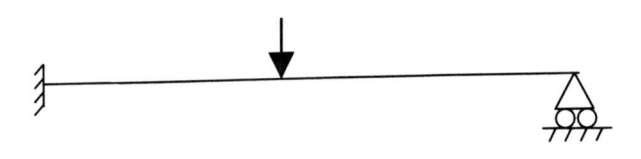

Figure 1.2: Propped Cantilever

Suppose again that support at B is lost. The beam can remain **stable** in this new configuration *only* if the beam can resist the cantilever moments and shears, and the support at A can resist the cantilever moment and shear, i.e., **if it is strong enough**. Only sufficient strength is required so that warning of collapse is given. The structure should be stable after an accident for a long enough time to evacuate the area. Accident safety factors apply, e.g., Eurocode EC2 (BS 2005a) requires ($DL+0.5IL$) with γ_m = 1.0 and 1.2 (for steel reinforcement and concrete respectively), where *DL* is the dead load and *IL* the imposed load. Serviceability criteria (e.g., deflections, cracking) do not apply.

Unfortunately there are some disadvantages of statically indeterminate structures. All of the following result in stresses in the structure (causing the conventional elastic solution to err).

- Boundary conditions: e.g., differential settlements of supports;
- Temperature changes;
- Shrinkage or creep of concrete structures;
- Members forced to fit/fabrication errors.

According to Heyman (2008) a steel propped cantilever's support would only have to settle by 1/1000 of the span to cause an increase in the bending moment at the fixed end of 15%. The need for ductility of both the material and structure is illustrated by Hambly's Paradox in the Appendix 1. This ductility allows the use of the elastic theory to obtain a safe solution (that it is safe is demonstrated by the Lower Bound Theorem which is proven in Heyman (2008) too).

Several surveys of actual stresses experienced by real structures have been carried out over the years, (e.g., see Heyman (1999) and Petroski (1992)). In the latter reference Petroski mentions some results of strain gauge measurements on electric pylon structures: the elastic stresses were typically only within 60% of the actual measured values. This can be regarded as a fairly typical figure. Clearly then the conventional elastic solution is not likely to represent the state of stress in the structure.

1.2 Ductile and Brittle

Of course we would prefer if the structures we design have no brittle elements. However, this is not possible. Any practical structure is likely to have brittleness incorporated in it. Brittleness can come from the nature of the material itself (e.g., a concrete tied column or shear of concrete), or it may come from the structural use of the member whether the material is ductile or not (e.g., buckling of steel or many connections).

The way to deal with these so the overall behavior remains ductile is simply to ensure that the brittle modes are given a higher Factor of Safety than the ductile modes. Since the load is increased from zero, the ductile modes are encountered first. The Factor of Safety for brittle modes must be significantly larger than that for ductile as our analysis is conventionally based on linear elasticity and 'perfect construction'. In addition, the actual load level at failure is usually more difficult to predict if the mode is brittle that when the mode is ductile (e.g., flexure can be predicted to about 5% accuracy if the material strengths are well known, but shear can only be predicted to perhaps 15-20% accuracy or so, especially if no links are used, as it depends on the tensile strength of the concrete).

Some of the quizzes in the section on worked examples look at the Safety Factors adopted by EC2. It is clear that it, along with other previous codes, adopts a 'strong-column, weak-beam' design philosophy but it is shown that its provisions relating to flexure and shear in RC require careful application.

1.3 Statical determinacy

When the structure is statically determinate there is no alternative load path to the supports. With each redundancy another possible load path is available. Thus the more redundancies the "safer" the structure can be made.

Mostly, the engineer has no choice in the matter of whether the structure statically determinate. The challenge is to make it safe in the event of an accident which removes the primary structure.

(There are two common exceptions to the general rule that statically determinate structures are to be avoided: steel beams and precast concrete beams. Here the designer explicitly provides an alternative load path. If the steel or concrete beam is properly tied to the supports, then catenary action (defined below) acts as an alternative means of resisting the load once the beam has failed.)

After an accident how can a reinforced concrete beam carry load?

1. Beam action: remaining RC beam tries to support load spanning to its supports.
2. Catenary action: concrete is assumed to have failed; only the resistance of the reinforcement remains.

Clearly, beam action is likely to be effective only for small accidents. Catenary action provides us with a much more likely load resisting mechanism. Consider the following example: The two-span continuous beam below in Figure 1.3 is designed to resist a uniformly distributed load of w ($=1.35DL+1.5IL$ according to EC2).

Suppose the central support is removed (say by an explosion or vehicle impact). Then to have a chance of survival, the reinforcement in the bottom of the beam must be lapped over the central support. The beam now (spanning $2L$) should be capable of resisting an accident load w_a ($= DL+0.5IL$ and material factors for design are to be taken as 1.2 for concrete and 1.0 for steel in EC2).

After the accident, the shape taken up is referred to as a "catenary". See Figure 1.4.

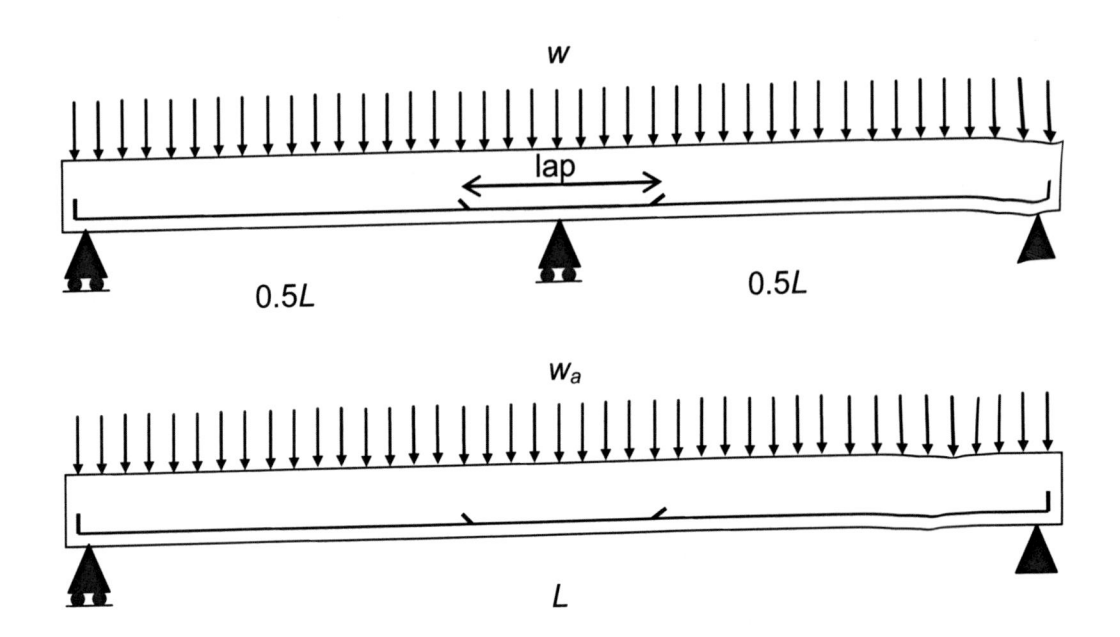

Figure 1.3: Removal of central support of beam.

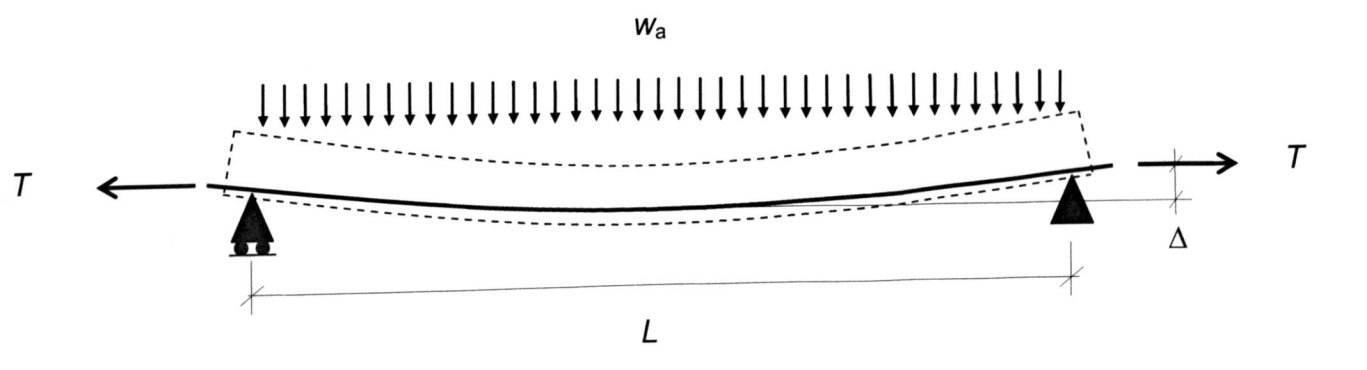

Figure 1.4: Catenary

The force T is $w_a L^2/8\Delta$ and is equal to $A_s f_{yk}$ where A_s is the area of reinforcement and f_{yk} its characteristic yield.

The tension force *T* grows from zero as the deflection increases. It becomes significant for deflections of about 0.5 x member depth. A capacity of at least 20 kN/m is required for *T* in EC2. Catenary action develops at deflections of about 0.1xL (Δ = 0.1L verified by experiment, see Elliott 1996). Clearly the detailing of the steel is crucial to allow the catenary action to develop.

1.4 Code approach

It is now appropriate to discuss the code's approach to ensuring Safety and Progressive Collapse Resistance. (EC2 is the code referred to in this section.) The code philosophy can be summed up as follows: *No structure can be expected to be resistant to excessive loads or forces that could arise due to an extreme cause, but should not be damaged to an extent disproportionate to the original cause.*

The design procedure is as follows: Code values of loading are adopted as "working loads" (assumed to be at their characteristic values).

1. Elastic analysis is carried out using these loads, factored by a safety factor, to determine ultimate values of moment and shear forces/reactions. Redistribution to allow for limited plasticity may be allowed for beams/slabs.

2. Members are designed to resist these ultimate values.

The Safety Factor (SF) is defined as: Ultimate Load/Working Load.

- Load factors for dead and live load are γ_f = 1.35 and 1.5 respectively. Thus Ultimate load = 1.35DL + 1.5IL;

- Steel γ_m = 1.15, Concrete γ_m = 1.5

- Assuming $DL \approx IL$

- Ductile (e.g., under reinforced Flexure): 1.425x1.15 = **1.64**.

- Brittle (e.g., over reinforced Flexure, 'tied' column): 1.425x1.5 = **2.14**.

The load effects factor, γ_f allows for:

- (i) uncertainties in applied load;
- (ii) errors in the analytical methods;
- (iii) effect of construction tolerance;
- (iv) allows simple methods to be used to assess serviceability.

The material effects factor, γ_m allows for:

- (i) variability of strength properties;
- (ii) difference between site and laboratory strengths;
- (iii) accuracy of methods used to determine strengths;
- (iv) variations in member geometry.

EC2 gives three alternative methods to ensure progressive collapse resistance.

1. "Ties": a "deemed to satisfy" method. This method is one usually adopted in practice.

2. "Alternative Load Path": Explicitly consider removal of <u>one</u> column on any storey and ensure collapsed area not large (UK building regulations (Great Britain 2004) suggest it must be less than the lesser of 15% or 70 m² floor).

3. "Key element": Design <u>all</u> members as key elements, using an ultimate load of 34 kN/m².

As solutions 2 and 3 are rarely adopted in practice, discussion will be confined to 1.

1.5 Tied solution

Reinforcement is provided so that the following ties can form in the horizontal direction. See Figure 1.5.

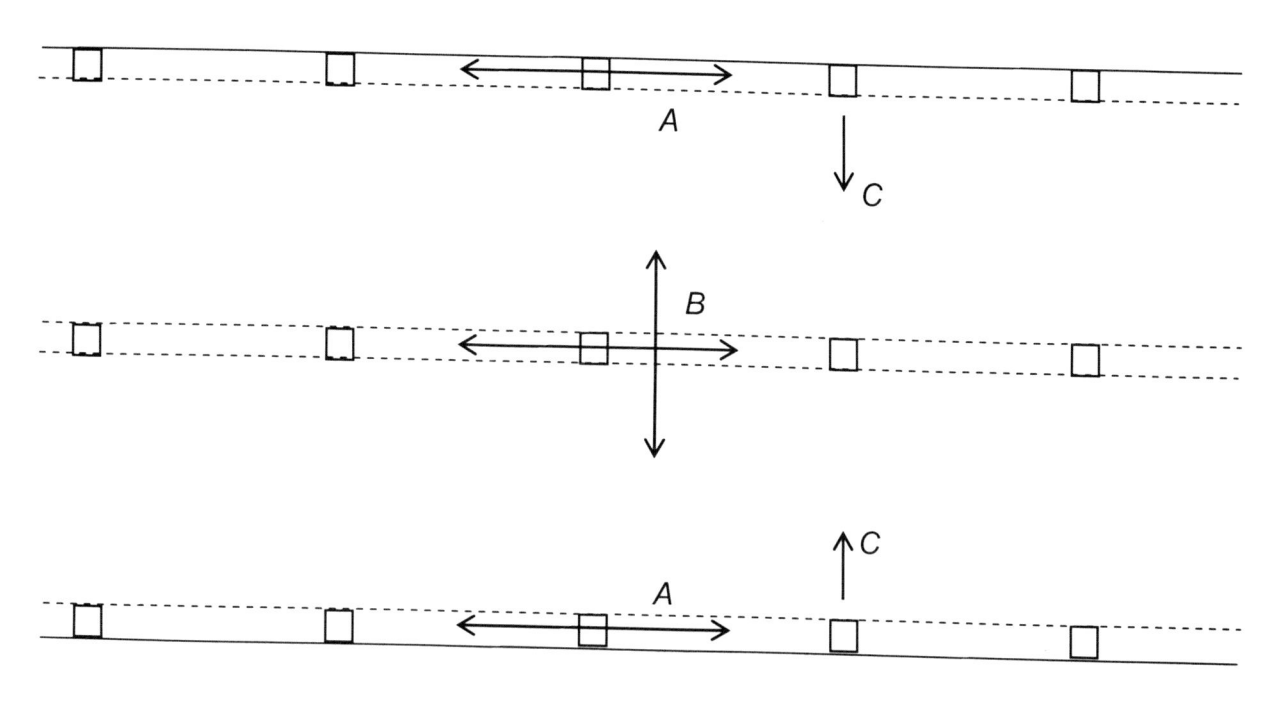

B = Internal tie; A = peripheral tie; C = Column/Wall tie

Figure 1.5: Tied Solution-horizontal ties

The purpose of these ties is of course to allow catenary action, as well as tying the columns into the building and tying the building together longitudinally. The entire reinforcement, originally provided for bending and shear, is available as it is assumed that the concrete has failed.

In addition vertical ties are required (formed by the column reinforcement which is available as the column is now being used as a tension element). These ties are there to ensure that the structure redistributes the load in the event of a loss-of-column accident at a lower level, such as the following (see Figure 1.6):

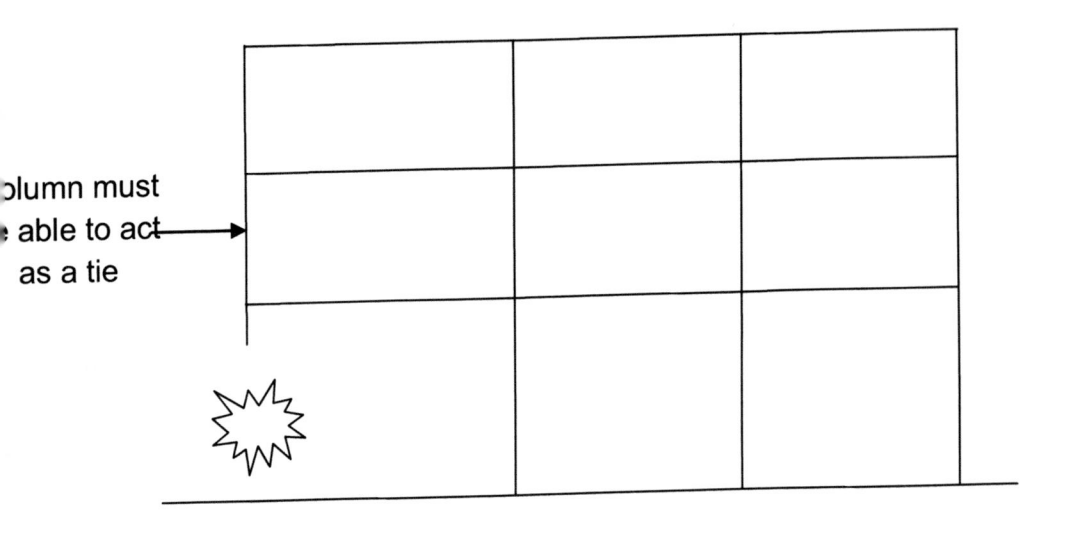

column must
able to act ⟶
as a tie

Figure 1.6: Tied Solution-vertical ties

Chapter 2: Gravity

"Good judgment is usually the result of experience, and experience is frequently the result of bad judgment"

Barry LePatner (Construction lawyer) in LePatner et al. (1982)

This chapter consists of failures where the only significant force acting at the time of failure was gravity. These are the most difficult failures to try and "justify" as gravity has always to be resisted and is a force of predictable value.

Four case studies are presented:

- Walkways at the Hyatt Regency Hotel, Kansas, 1981;
- Hotel New World, Singapore, 1986;
- Sampoong Department Store, Seoul, 1995;
- Paris Airport, France, 2004.

2.1 Walkways at Hyatt Regency Hotel, Kansas, USA, 1981.

[Ross (1984), Dowling et al. (1988), Kaminetzky (1991), Levy and Salvadori (1992), Petroski (1992), Petroski (1994), Schlager (1994), Feld and Carper (1997), Moncarz (2000), Wearne (2000), Jennings (2004), Fleddermann (2004), Delatte (2008), Gerstein (2008)]

There were three steel and concrete walkways spanning the atrium of the hotel lobby which connected a function block to the accommodation tower (35 storeys). The walkways were hung from the roof of the atrium using steel rods. A single walkway was at level 3. The other two walkways were placed directly above one another at levels 2 and 4. The arrangement of the latter two walkways is shown in Figure 2.1.

Each walkway consisted of steel beams running along the edges and supporting a concrete slab. The steel edge beams were supported at each hanger position by a pair of channels welded together, forming a 'box-beam'.

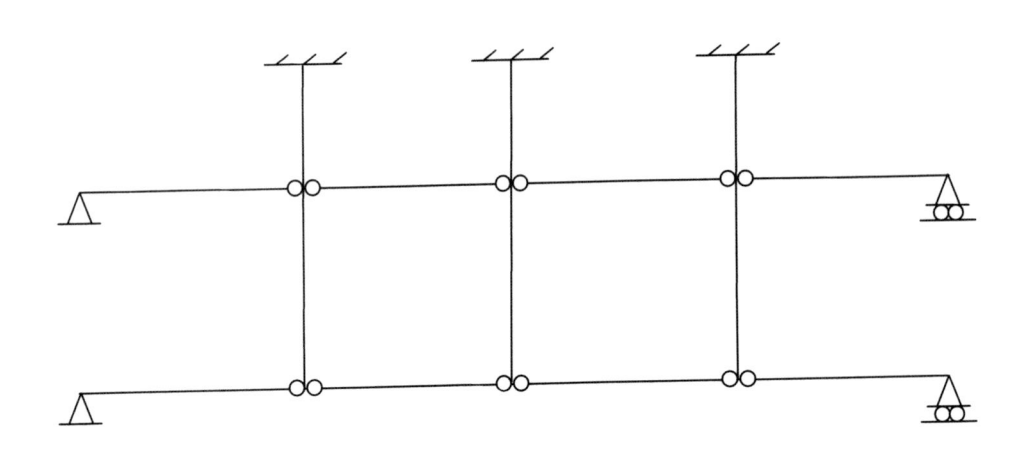

Figure 2.1: Structural model of walkways

Collapse

One evening, about one year after the opening of the hotel, a dance competition was being held in the atrium. People stood on the walkways, suspended over the atrium, for a better view. Suddenly there was a loud cracking sound and two of the walkways fell to the floor (4th and 2nd floor walkways). The hangers remained attached to the atrium roof. Clearly the connection had failed. The collapse killed 114 people.

Investigation

Comparison of the as-built connection and the as-designed connection demonstrated that the details were not the same. Instead of the intended continuous rod, the as-built construction included two rods. These are shown in Figure 2.2 below.

The as-built detail <u>doubled</u> the load on the nut at the end of the upper hanger rod, causing the upper rod to simply pull through the channels. The Engineer was aware of the change (he stamped "approved" on the fabricators drawings, which showed the revised detail), although perhaps unaware of the implications. The fabricator designed all connections elsewhere on the project, although apparently not this one. According to the engineer <u>the intention</u> was to have the fabricator design this connection too.

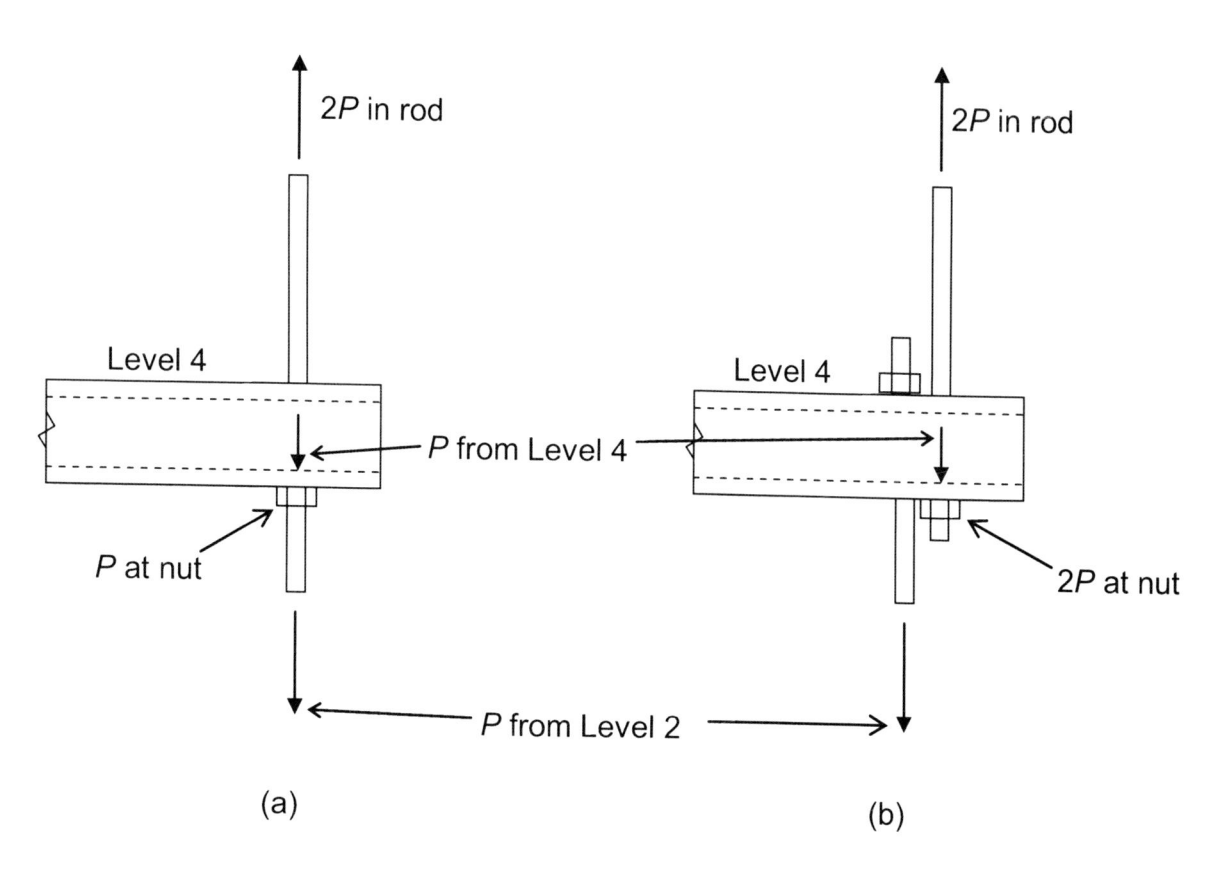

Figure 2.2: (a) As-designed connection detail; (b) As-built connection detail

Tests carried out on the remaining intact walkway (3rd floor) demonstrated that the connection only had 60% of the code specified capacity. Thus even the "as-designed" connection detail was not strong enough. The behavior of a similar connection is analyzed in Heyman (1982). This analysis showed the forces on the welds between the channels and the need for plates at the ends of the box beam (in addition to any at the bearing). The strength of any weld between the flanges of the channels is likely to be suspect if it ground flush with the flanges, as it was as in this case.

Lessons

- The engineer is responsible for the connection design even though the actual work may be carried out by others (in this case the steelwork fabricator). Although this case concerned steelwork the principle applies to any contractor-designed item, e.g., post-tensioned concrete.

- Poor communication between the designer and the fabricator meant the design of critical connections was neglected, resulting in nobody designing the critical rod/box connection carefully.

- Connection failures often occur suddenly and without warning and result in the loss of the entire structure if the structure is statically determinate as was the case with the walkways here. (A general principal is to avoid statically determinate structures where possible unless sufficient collapse resistance can be added, e.g., catenary action, see introduction). The factor of safety may need to be increased for connection design.

- Simple changes to the connection detail, e.g., putting a bearing plate between the nut and the box beam formed by the channels, could have improved the load carrying capacity significantly.

- The main reason the fabricator made the change was to improve the buildability of the design (there was no guidance as to how a nut could be placed approximately mid-height of a rod!). So ensure at least one method of construction is possible and clear to all.

"it was the job of the engineering company to create a process that trapped slips, engineering errors, and oversights".

Marc Gerstein (Author of Flirting with Disaster 2008).

2.2   Hotel New World, Singapore, 1986.

[Thean (1987), Hulme et al. (1993), Hulme et al. (1994)]

The building was a six storey RC framed building plus a basement car park. It consisted of RC beam and slab floors. The foundations were RC driven piles. Partitions were brickwork. It was constructed in 1971. See Figure 2.3.

Collapse

On the night before the collapse wide diagonal cracks appeared in some columns. The building collapsed completely and fell straight downwards, with no sideward movement. The owner and 32 others were killed in the collapse.

Investigation

From 1974 onwards there were some reports of cracks in columns or floors, or doors becoming stuck. Typically the owner ignored the reports or made superficial repairs. No engineer ever inspected the building.

As well as preparing engineering drawings the draftsman also prepared all structural calculations. It seems no supervision or checking by any qualified person was undertaken.

The prospective owner of the structure engaged the contractor and supervised him (there was no other on-site supervision). The owner had no engineering training. No proper material tests were carried out during construction (e.g., concrete cube tests, tension tests on steel, soil tests).

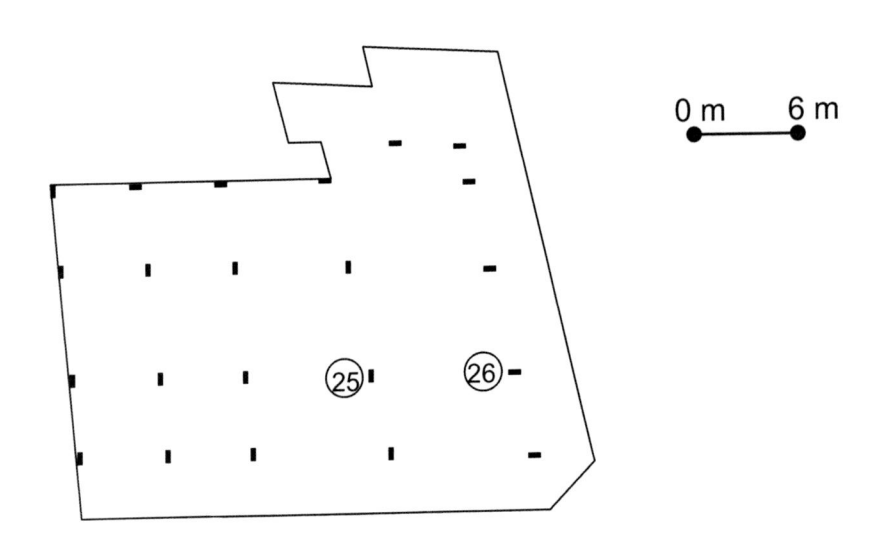

Figure 2.3: Plan of New World building

No evidence of corroding steel or chemical attack of concrete was found. The collapse is believed to have started in columns 25 and 26. The reason for the collapse was deterioration of the concrete strength due to the propagation of microcracks. This phenomenon is called "static fatigue". It is caused by long-term overstress of concrete. The strength of a cylinder of concrete that is familiar to us all is the short-term strength. If the compression stress is maintained at a high level for a long time (say $> 0.7f_{cu}$), microcracks grow parallel to the loading (i.e., vertical in the case of a vertically loaded cylinder) and eventually cause the cylinder to fail. See Figure 2.4 below. Proper design means that the service value of the stress is unlikely to exceed $0.33f_{cu}$ and so static fatigue is never a worry. In this case poor design meant the stresses were very high for 15 years before collapse. Careful checking of the calculations after the collapse revealed the elementary mistake of forgetting to include the self-weight of the concrete.

Effectively there was poor redundancy too as when one column started to fail the load could not be redistributed to neighbouring columns which were close to failure too.

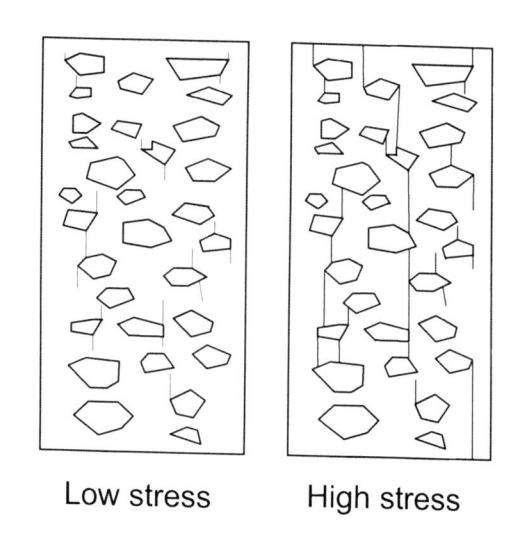

Figure 2.4: Concrete under low and high compressive stress

Lessons

- Structural design should be undertaken by appropriately qualified and experienced people.

- The design must be checked and the construction supervised.

- Sufficient tests should be done on the construction materials to ensure the quality of the work.

- Professional advice should be obtained so that any cracks or sagging in the completed structure could be properly interpreted.

Additional information

One reason ancient concrete structures such as the Pantheon (constructed circa 123 AD) have survived to the present day is because of the low stresses used.

2.3 Sampoong Department Store, Seoul, Korea, 1995.

[Gardner (2002), Wearne (2000), Delatte (2008)]

[handwritten: Reinforced concrete to strengthen - plates incorporate in tension]

The building was a nine-storey high *in situ* RC structure consisting of five upper floors and four basement levels. Foundations were piled to rock. A schematic of the North wing is shown in Figure 2.5 below. There were two wings (North and South) separated by atrium. Circular RC columns were 10.8 m apart. The floors were typically 300 mm thick flat slabs (drops increased overall thickness to 450 mm). Stair cores braced the building. The building was completed and opened in December 1989.

Collapse

In the months before the collapse there were many cracks, leaks and even bangs, together with problems with the air-con. On the day of the collapse, 29 June, structural engineers examined the building and told the owner the building was unsafe. Company executives who met that afternoon decided otherwise but quickly left themselves. At 17:55 the entire North wing suddenly collapsed, except for the stair cores at the four corners (see Figure 2.6). Nearly 500 people were killed.

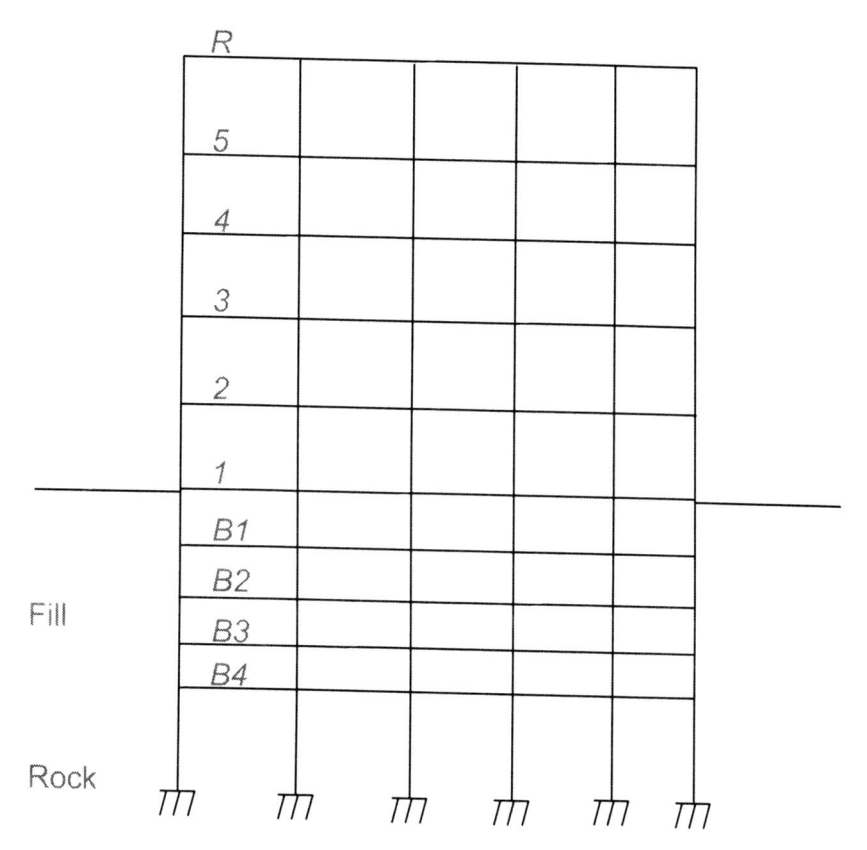

Figure 2.5: Schematic of the North wing

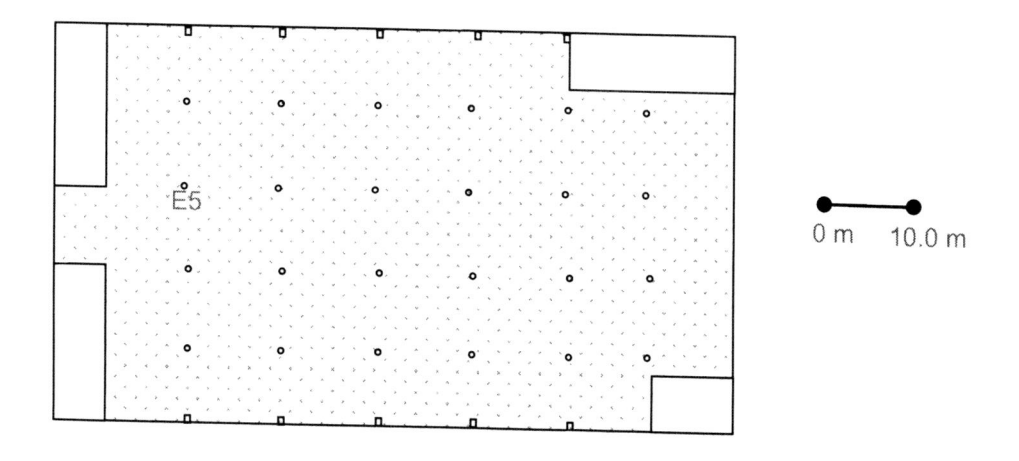

Figure 2.6: Extent of collapse of North wing (shaded)

Investigation

No extreme weather or seismic activity at the time was reported. According to witnesses collapse started at an interior column at level 5 (E5). Calculations suggest that it may have initiated at the same column on the roof level too. In any case, collapse quickly progressed to the lowest level.

The building was designed as an office. However during the construction the owner decided its use was to be changed to be an up-market department store.

Some of the changes were:

- Level 5's use was changed to traditional Korean restaurant (requiring a new concrete false floor-resulting in 35% increase in dead load).

- The air-con system was upgraded, e.g., three new water towers, weighing up to 30 tonnes each, were added to the roof. The original design assumed any additional load would be less than 1 kPa. The roof slab was thickened by using RC as result. (The towers were later moved by sliding from one end of roof to the other, as a result of neighbours' complaints about noise. Unknown damage occurred.) Many new walls from 4-5 were added, further increasing dead load.

- Many new floor penetrations were made, e.g., to add escalators.

- More stringent fire-related requirements applied, e.g., new fire doors were added resulting in removal of part of columns.

The design used the then current Korean code which is similar to ACI-318-83. The original <u>as-designed</u> structure generally satisfied this code (except for the minor error of neglecting drop self-weight.) In the <u>as-built</u> structure, however, the failure of the connection at E5 on level 5 was predicted by BS8110-85 but not by ACI-318. ACI-318 is known to give poor predictions of punching strength when the percentage of flexural reinforcement is low (< 0.5%) and for thick slabs. Sampoong had both.

ACI-318-**89** requires bottom bars ("hanger bars") through connection as "integrity steel" (See Figure 2.7). But ACI-318-**83** does not. So no bottom bars were provided through the column. Thus when punching failure began (probably on the roof), the collapse progressed to the basement. Thus the layout of post-punching reinforcement for a flat slab floor should be as follows (see Figure 2.8):

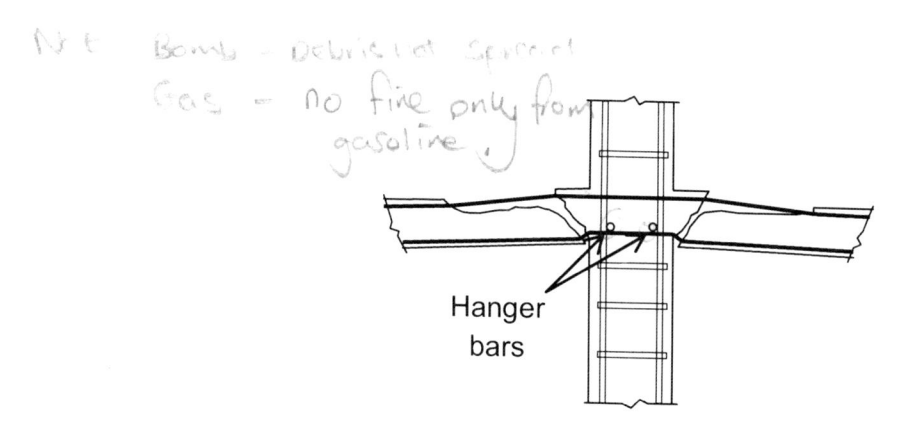

Figure 2.7: Post-punching reinforcement

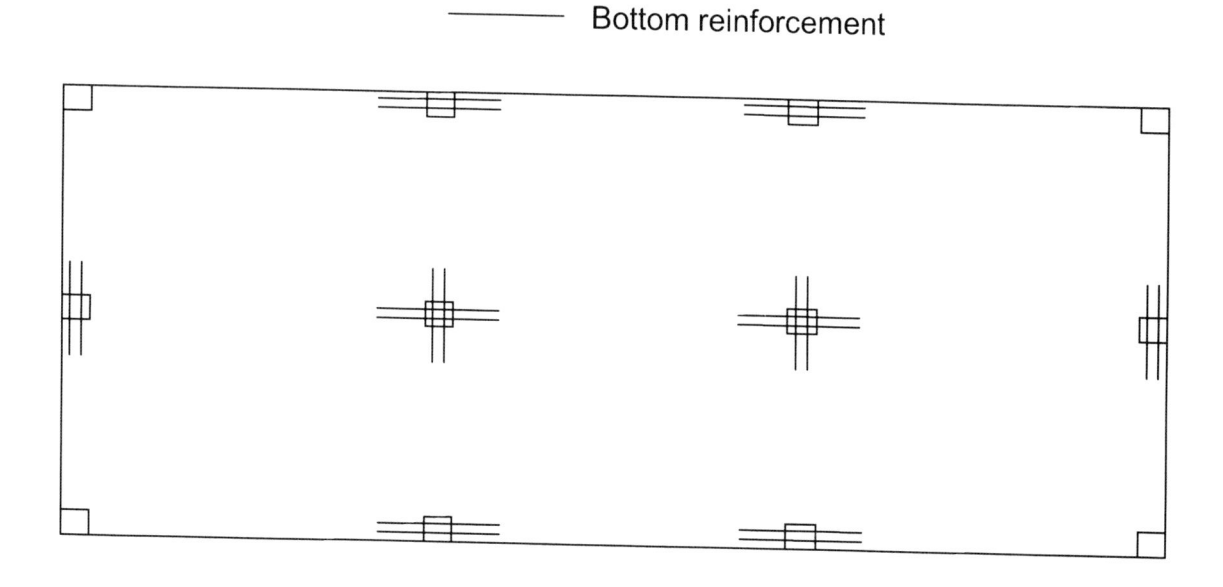

Figure 2.8: Layout of bottom reinforcement

Lessons

- Post-punching resistance of a flat slab structure is low.
- Punching failure at an upper floor can trigger a progressive collapse to the lowest floor. This tells us that preventing the floor from falling is critical.
- Provide bottom reinforcement through each column cage in a flat slab structure to provide progressive collapse resistance.
- Code predictions of punching resistance can be quite different; thus treat any predictions of shear related phenomena with caution.
- Symptoms of structural distress evident, e.g., cracks, leaks, noises, and warning from structural engineers.
- The owner has responsibility to act on warnings to protect public.
- This failure reminds us as designers that there is much that is out of our control and so to be cautious.

Additional Information

Some research has been carried out on post-punching behavior (e.g., Hawkins & Mitchell (1979), Regan (1981)). It is recommended by Hawkins & Mitchell (1979) that the capacity of a single bar is taken as $0.5A_s f_y$. Since each bar acts in "double shear" the capacity of a bar is $A_s f_y$. Thus the number of bars required = Reaction/$A_s f_y$. A minimum of four bars should be provided, i.e. a cruciform shape on plan.

Quiz 21 gives an example of the provision of bottom bars.

Other advantages of bottom bars:

- Cheap (usually accomplished by extending main span reinforcement into support).
- Help reduce deflection increase due to creep.

Perhaps surprisingly, provisions for bottom bars were only slowly introduced into codes of practice, despite clear evidence for their need. For example, the following codes require them:

- American code, ACI-318, since 1989.
- Canadian code, CSA A23.3, since 1994.
- British code, BS8110, since 2005 (code now withdrawn).
- European code, EC2, since publication.

However, EC2 does not control the diameter or length of the bars.

Note that in the common *in situ* beam-and-slab floor system it is important to provide this bottom reinforcement in the beam (top reinforcement helps too due to the action of the links) to prevent any shear failure in the beam from developing into a progressive collapse. However care should be taken at the end support; reinforcement in the edge beam (i.e., perpendicular to the main beams) can help avoiding the triggering a progressive collapse.

2.4 Terminal 2, Paris Airport (Charles de Gaulle), 2004.

[Wood (2005), Petroski (2006), Reina (2005), Berthier (2006)]

The terminal building was a tube-shaped building known as the "jetty". It was formed by an RC barrel vault, with window openings. This shell was then covered by non-structural glass and aluminium. The terminal was opened to the public in June 2003.

Collapse

Warning was given about 90 minutes before the collapse by the falling of a large piece of concrete from the ceiling. Police evacuated the area. Shortly afterwards a portion (24 m) of the tube building collapsed. The collapse centered on where three walkways entered the tube on its north side. The collapse killed 4 people.

Investigation

The enquiry into the collapse blamed the design.

The tube structure was complex. The basic structure consisted of a series of elliptical reinforced concrete arches 300 mm thick each spanning 26.2 m. As well as having conventional bar reinforcement, each arch was connected to structural steel external trusses. The shell concrete was monolithic with longitudinal edge beam which was in turn tied to and supported on columns.

Each arch rib was 4 m wide and constructed in 3 sections, which were joined on site. There was no connection between each arch rib. See Figure 2.9.

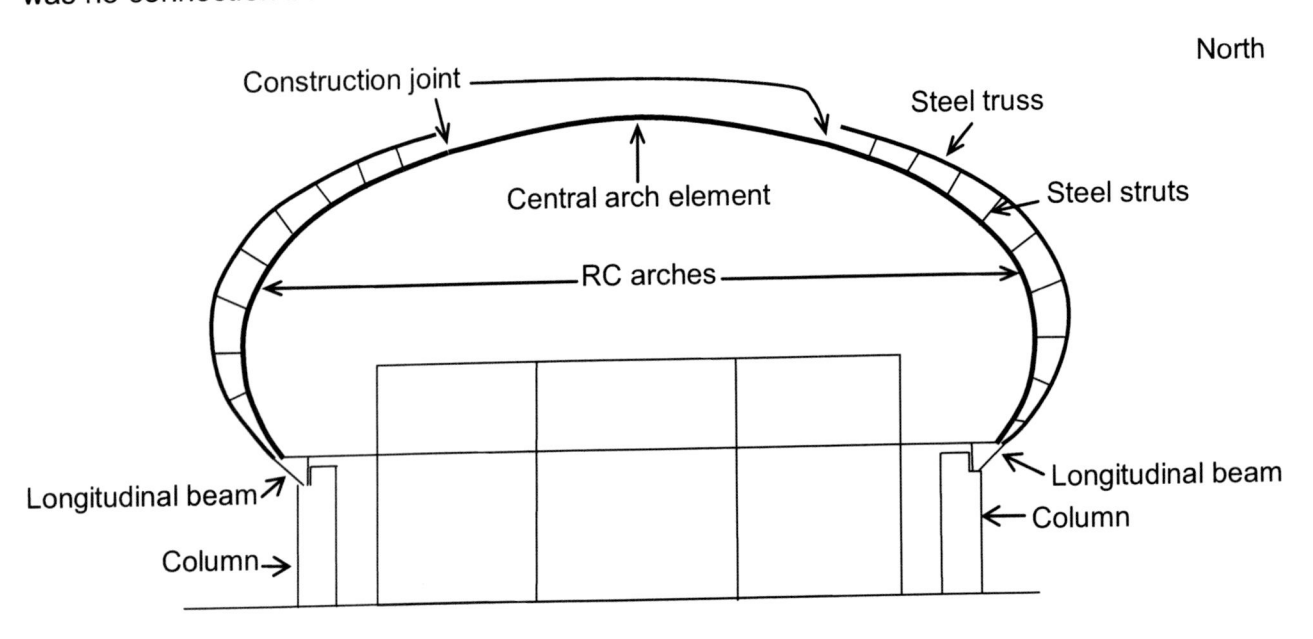

Figure 2.9: Cross-section through "jetty" building

At the collapse zone the structure was further complicated by the presence of three footbridges on the north side of the tube.

The collapse was the result of two nearly simultaneous events:

1. Punching of some struts through concrete shell and flexural failure on North side. This destroyed the composite action between the steel truss and the concrete shell.
2. Sudden push to South from folding shell resulted in failure of longitudinal beam and fracture of column tie on South side.

The enquiry noted that the strength of the structure had probably deteriorated since construction (completed over 2 yrs previously). This deterioration was probably the result of differential movements of concrete and external steel as a result of temperature, creep and moisture movements of the concrete. Evidently it caused the progressive development of cracking in local areas of high punching stress.

The precise positioning of strut plates, which were intended to be recessed inside concrete, was sensitive to construction tolerances resulting in unnecessarily high punching stresses which caused cracking. This cracking meant that there was a need for steel reinforcement. There was none so the sudden collapse was clearly the result of brittle failure of this unreinforced concrete. The arches were essentially independent 4 m wide sections, so the overall redundancy was poor.

The innovative nature of the building was recognized during design: three independent studies were carried out on the design before construction (as required by Eurocodes for "innovative" structures). None identified the possible need for punching reinforcement.

Lessons

- In some structures (e.g. shells) degradation of structural strength over time can happen as a result of cracking due to temperature differences, shrinkage or creep. Deflections should be monitored preferably for at least the first few years (Chilton 2000).

- A low initial reserve of strength (resulting from high percentage of load, approximately two-thirds in this case, being permanent) is sometimes likely if code recommended factors of safety are blindly implemented. As the majority of the load was permanent the consequences of any deviations from the analysis were higher.

- Consequences of collapse were high: thus a higher factor of safety than the code recommended would be appropriate.

- Redundancy was poor. There was little load shedding ability from the arches. Also an arch does not have good redundancy (being reinforced primarily for axial load).

- Peer-review is essential to expose modelling errors that are not clear to designers. This advice was followed here, in fact there were three checkers, but it is possible that a single checker would have been more effective than multiple checkers!

- Careful when concrete interacts with structural steel in bending, as the prediction of behavior is especially difficult.

- Pre-conceived architectural solutions are rarely appropriate for shell structures (see also Mark 1990). The requirement to use a non-funicular shape led to large moments and left the shell prone to secondary effects e.g., change in shape and subsequent increase in moments due to creep which changed the interaction between the steelwork and concrete.

- The structure was a complex one involving the interaction of structural steel and concrete. Linear elastic analysis is very sensitive to boundary conditions (differential movements, lack of fit, etc). Ductility is required to overcome this "deficiency" (see Appendix 1). Thus a high factor of safety against brittle modes of failure (punching in this case) is needed.

Chapter 3: Computers and Modelling

"To err is human; to really mess things up, you need a computer"

(Anonymous)

This chapter consists of failures where mistakes in the structural analysis contributed to the collapse.

Four case studies are presented:

- Hartford Civic Centre, USA, 1978.
- Sleipner A, Norway, 1991.
- Ramsgate Walkway, UK, 1994.
- Compassvale School, Singapore, 1999.

3.1 Hartford Civic Centre, USA, 1978.

[Ross (1984), Kaminetzky (1991), Ferguson (1992), Levy & Salvadori (1992), Petroski (1992), Murta-Smith (1993), Petroski (1994), Schlager (1994), Shepherd & Frost (1995), Feld & Carper (1997), Various (1998), Wearne (2000), MacLeod (2005), Delatte (2008), Ratay (2010)]

The roof of this arena was 110 m x 90 m and 6.4 m deep and was supported by four columns. It was constructed in 1973. It consisted of a space frame/truss which was constructed using angles arranged to form a cross. Members were arranged into a repeating inverted pyramid module (plan dimensions 9.15 m x 9.15 m) as shown in Figure

3.1. The roof purlins were not placed directly on the top chord of the truss, but on short posts, in order to achieve a fall in roof (this avoids having to camber truss and directly loading the top chords).

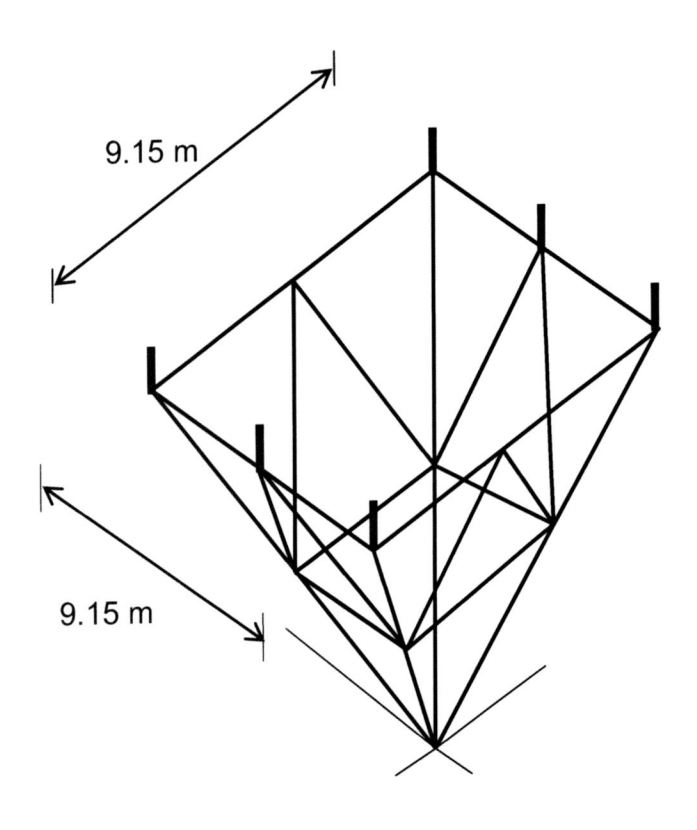

Figure 3.1: Inverted pyramid module

A linear elastic analysis was by means of the computer program (STRUDL). Space frames were seldom built before computer programs were available.

The truss was assembled on the ground and jacked up the four columns into place. The compression and tension chords consisted of four angles connected together in a cruciform shape, while the diagonals consisted of single angles. This meant that the members of the truss were not connected to each other concentrically.

Collapse

Early one Sunday morning, about five years after construction, the arena experienced the worst snow storm since its construction. The roof suddenly collapsed to the floor of the arena. Only five hours before the collapse there were 5,000 spectators in the stadium to watch a hockey game. Luckily, all had left the stadium, so there were no injuries.

Investigation

The snow load at the time was estimated to be about 1 kN/m^2 (corresponding to 75% of design value). The structural model input into the computer ignored all eccentricities present in the connections. The actual dead load of the roof was 0.72 kN/m^2 heavier than in calculations (this meant there was 25% "extra" dead load).

Design calculations assumed each compression member in the top chord was restrained at about 4.58 m (= 9.15/2) centres by the diagonals. This was true of the members in the interior of the frame, but not those on the perimeter. The effective length for those members was 9.15 m. Thus the design calculations and the details were not consistent.

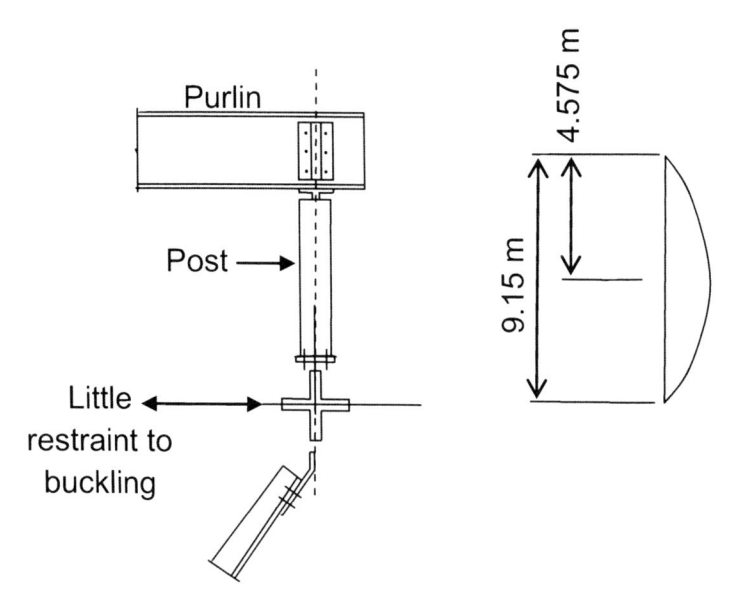

Figure 3.2: Structural configuration at perimeter

In addition, perimeter members were loaded with a post but not designed for bending. In the computer all load was applied at nodes. Figure 3.2 shows the structural arrangement at the perimeter.

The only on-site inspection was by the construction manager (not a structural engineer). He claimed, not unreasonably, that his only responsibility was to ensure roof was constructed in accordance with drawings. The designers had no contract for inspections. A photograph taken during construction shows obvious bowing of two members in top layer.

Later computer analysis also showed that failure of one compression member would progress into collapse of whole roof. This was mostly a consequence of the arrangement of the roof supports (Murta-Smith 1993). Thus although highly statically indeterminate, the roof had poor redundancy.

Unheeded warnings

1. Actual dead-load deflections were *twice* that predicted by the computer model. Engineers said large deflections were "to be expected".

2. On two occasions lay-people voiced concern about obvious deflections and were reassured that all was as expected.

3. Construction workers noticed large deflections.

4. Contractor installing perimeter fascia panels could not align the pre-punched holes.

Summary of the main mistakes

1. Failure to brace top members adequately.

2. Failure to allow for eccentricities in model.

3. Failure to include post loading.

4. Miscalculation of dead load.

Lessons

- Computers give the answer to a specific question. Ensuring that question is correct one, i.e., that the model represents reality is job of the engineer.

- Add redundancy where possible to avoid progressive collapse.

- The consequences of failure of a sports stadium roof are potentially severe: increase the factor of safety.

- Do not ignore warning signs.

- Eccentricity can be ignored, and often is, in a normal plane truss as it is small. But here eccentricities were too large to ignore.

- A cross not particularly good arrangement for a member taking compression (since this shape has a low radius of gyration). Thus resistance to compression and bending is poor. Torsional buckling is likely to be a risk for rolled sections (Trahair et al. 2008).

- The actual performance of the structure, e.g. measured deflection during construction, is perhaps the only indication the engineer has that the structural model is representative. This counter-checking should always be done, especially for complicated structures.

- Computer analysis may provide a false sense of security.

- Long-span structures in particular need careful inspection by qualified people during construction.

- Project peer review required.

- Failure due to instability is a brittle failure, and the buckling of one member can lead to collapse of the entire structure. Thus the Factor of Safety should be increased to reflect the uncertainty of the analysis.

Aftermath

The arena was rebuilt by 1980. The new roof is simpler: it is composed of two ordinary parallel vertical trusses sitting on the same four columns. Secondary trusses frame into these trusses at six locations. In addition, today, Connecticut is one of few states in USA that require peer review of certain buildings.

Additional Information

In 1979 the Gerald Ford Museum (Gerald Ford is a former US president) was being constructed with a space frame roof. Ford instructed that load-test was to be carried out to prove safety of the completed roof. Water was used to impose a load of 2 kN/m^2 on the roof. The resulting deflection was considered satisfactory (37 mm). The cost of the test was about 1.5% of the total building cost.

General notes:

Space frames are usually highly indeterminate structures. However this means that there is a greater chance that there exists a state of self-stress even before load is applied (think of all the members that had to be forced into place to fabricate the truss). So the conventional linear elastic analysis is unlikely to be very accurate. Thus a higher than usual factor of safety on the compression members is required, especially as these structures tend to cover a large space. In addition, the analysis should be checked using in situ measurements of deflection.

3.2   Sleipner A offshore platform, Norway, 1991.

[Collins et al. (1997), Petroski (1997), Various (1998), Kotsovos & Pavlovic (1999), MacLeod (2005)]

"Sleipner A" was a reinforced concrete gravity oil platform. In service, it was supposed to sit on the sea bed in 82 m deep water. Its four hollow legs (C1, C6, B3 and D3 below) housed drilling equipment, and oil was to be stored in the other 20 concrete cylindrical chambers which were sealed at both ends. The walls between the legs and the chambers were known as "tricells". Water could enter the tricell to subject the walls to the full head of water pressure. A plan of the base is shown in Figure 3.3.

Construction of the concrete structure was completed in a dry-dock in Norway.

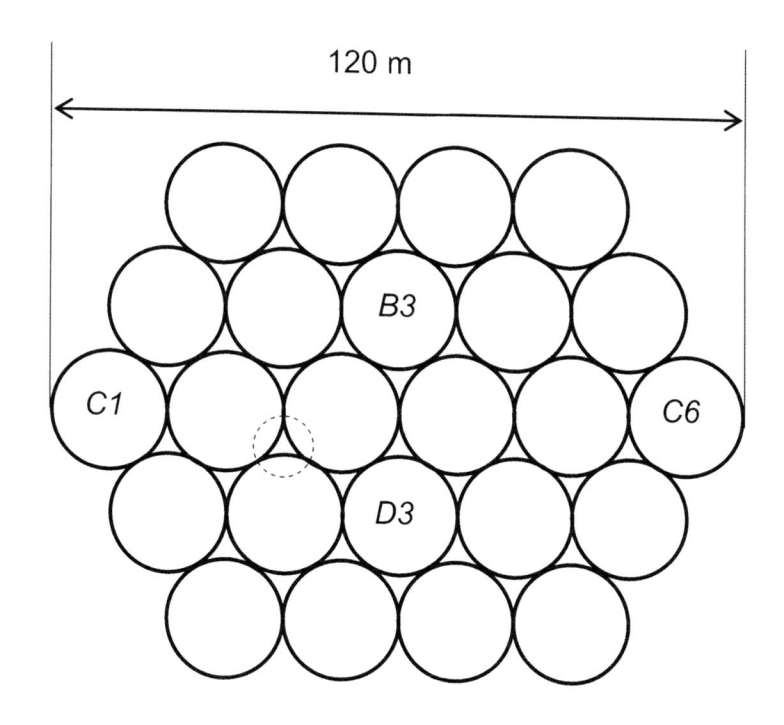

Figure 3.3: Plan through cells (typical tricell highlighted)

Collapse

The entire concrete structure was towed out to deep sea (water over 200 m) to be connected to the steel superstructure. The concrete structure was to be partially sunk so that it floated with its 4 legs protruding only a few meters above the water. The steel superstructure was then to be towed over the legs. The concrete structure was successfully sunk, but it couldn't be stopped from sinking. It sank to the bottom of the sea. The high pressures there caused it to implode. The estimated cost was US$300 million.

Investigation

The wall of tricell 23 probably suffered a shear failure. The design fault was traced to major mistakes in the analysis. A general finite analysis program (called NASTRAN) had been used for the analysis of the RC structure under the water pressure. The entire concrete structure was modeled using 3D ("brick") elements. The tricell wall was modeled using one element over the width of the wall. Unfortunately a single "brick" element is unable to model bending properly. In addition, considerable distortion of the elements was necessary for them to model the fillets at the junction of the walls. The two mistakes resulted in a 45% underestimate in the predicted shear force in the wall. Simple hand calculations using a fixed-ended beam could have detected this error but were never done. See Figure 3.4. Although the section was relatively heavily reinforced in flexure using top and bottom reinforcement, no shear reinforcement was present.

Kotsovos & Pavlovic (1999) showed that the point of contraflexure is likely to be a point of weakness. Each span had two. Without reinforcement (i.e., links) the tensile strength of the concrete is the only means of generating the 'internal support' required. For example, a two span beam and its bending moment diagram are shown below in Figure 3.5. The thick line represents the path of the compressions through the structure while the narrow line represents the reinforcement. Notice that each point of contraflexure is a region in direct tension (needed to equilibrate the vertical components of the inclined compressions). Vertical links must be provided here.

In each fixed-ended span of the tricell wall there were two points of contraflexure (at about 0.2xspan from each support). Thus only the tensile strength of the concrete was available to resist the tensions in that region.

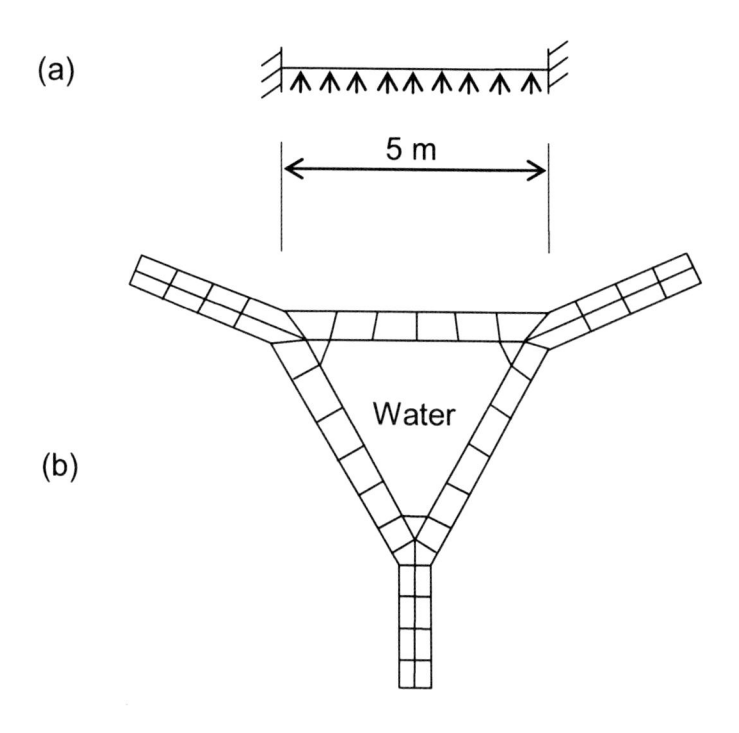

Figure 3.4: (a) Fixed ended beam checking model for tricell and

(b) Finite element mesh of a tricell

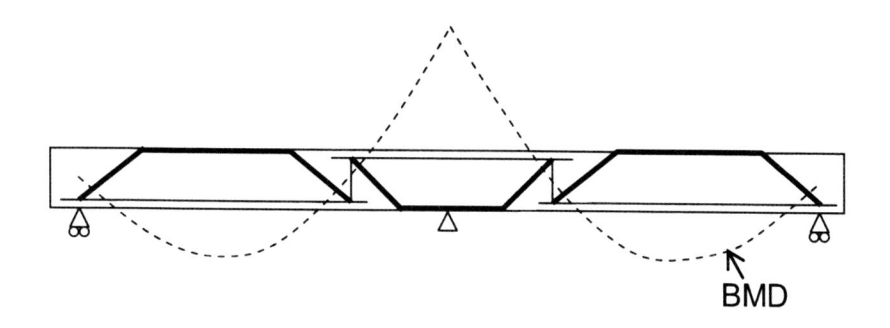

Figure 3.5: Two span continuous beam

Using a Non-linear Finite Element program (Collins et al. 1997) it has also been shown that the provisions of the ACI-318 code regarding shear combined with axial compressive force, are potentially unsafe.

Lessons

- Computer modeling, however complex, can usually be checked using hand calculations on a simplified model - a "checking model". They should always be done.

- For this element it is recommended that at least four elements are used across the width of a member to model bending properly. This advice applies to shell elements in general purpose finite element packages.

- Shear failures often occur suddenly and without warning and may result in the loss of the entire structure if no post-failure precautions have been included (i.e., alternative load paths/bottom reinforcement). In this case the resistance to water had to be maintained so any shear failure was catastrophic.

According to Collins et al. (1997) it was the "most expensive shear failure on record".

Suggested method for checking computer output (MacLeod 2005)

1. Validate the model (can the model solve the problem?)
2. Verify model represents the actual structure.
3. Check the input data.
4. Check the reactions.
5. Check the deflected shape looks correct.

6. Use simple load cases to check symmetry and effect of changing the major parameters.

7. Use a simplified model of the structure (a 'checking model') to estimate the results.

3.3   Ramsgate Walkway, Port of Ramsgate, UK, 1994.

[Chapman (1998), Various (1998)]

This case study concerns the failure of a walkway that linked a ship Terminal Building to a main floating structure. There were two levels of connection: an upper one for pedestrians and a lower one for vehicles. Tidal movement meant the floating structure rose and fell. The upper walkway, the one of concern here, was a covered steel footbridge. The arrangement is shown in Figure 3.6. It was constructed in May 1994.

The client ("Port Ramsgate") appointed an experienced and well-regarded contractor to carry out work as design-and-build contract. Design, fabrication and construction were certified by Lloyd's Register.

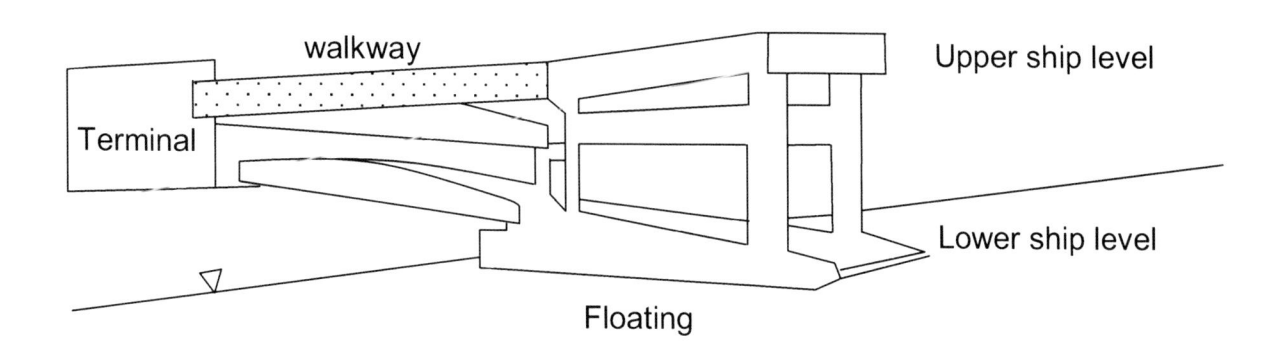

Figure 3.6: Ship to Terminal connection

Collapse:

On 14 September 1994, during the loading of a ship the walkway suddenly, and without warning, collapsed. The seaward-end fell 10 m onto the floating structure and the shoreward-end remained in place. Six died and seven were injured. The walkway is shown in Figure 3.7.

Investigation:

The walkway structure was inspected (visually) one month before the collapse. It was analyzed as a simply supported beam spanning 33.7 m. The structure consisted of box section members connected by flat bar cross-members to form a truss.

The "pinned" bearing was to the floating structure (sea-end) end and the "sliding" bearing was at terminal end (shore-end). The left-hand bearing at the sea-end was also a sliding bearing. The sliding bearings had no provision to resist uplift. The roof, lower sides and floor were clad with 6 mm steel plate; the upper sides were clad with glass.

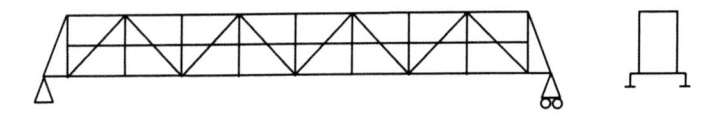

Figure 3.7: Walkway elevation and section

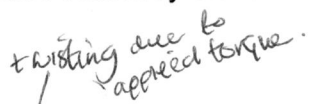

twisting due to applied torque.

Thus structure was very stiff torsionally. However, torsional movements were not allowed by the bearing system. Rolling of the structure could only be accommodated by a bearing lifting from its seating. A proof load test (consisting of the application of the full static live load) had been performed by Lloyds before walkway entered service.

The trigger for the collapse was the fatigue failure of the welded connection between the pin and walkway. Two heavy vehicles had just crossed the lower linkway onto the floating structure. They caused the floating structure to roll and this triggered the collapse of the walkway.

[handwritten: microscopic cracks caused by cyclic loading.]

The prime cause of the collapse was a very basic modelling error by the designer. Structurally the bearing assembly was a cantilever. The designer had neglected the presence of a bending moment at the root of the pin. Figure 3.8 shows the bearing that failed first. The weld was designed to resist the shear force only.

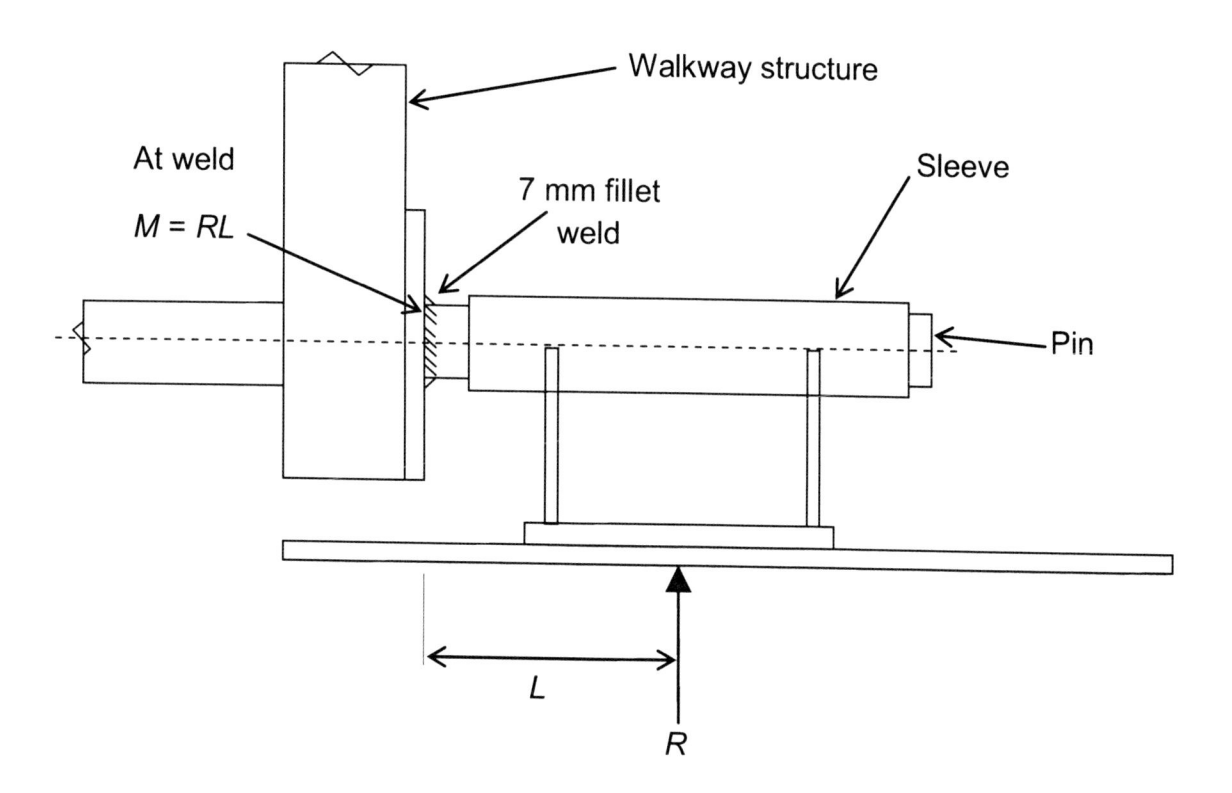

[handwritten: beam anchored at one end]

Figure 3.8: Arrangement at bearing

In addition, the designer had assumed the reaction at all 4 support points was equal. Thus they assumed the reactions were half what they could be as a result of uplift of one bearing.

The design calculations and drawings were checked and approved by Lloyds Register. The checkers made exactly the same modelling errors as the original designers. The court found the designer/contractor, checker as well as the client liable, and fined them £1 million, £500,000, and £200,000 respectively. Mr. Justice Clarke, High Court Judge noted that the: *"Jury convicted Port Ramsgate [Client] on basis that it failed to take steps to fit a fail-safe system, such as chains, in order to ensure that the walkway did not collapse"*.

Lessons:

- In this case the logical choice for the design was a statically determinate support system because of the large vertical support movements expected. However, even in such a case failure of the primary system need not mean total collapse. For example, chains could have been used to ensure alternative load paths.

- Do not rely on a checker to spot mistakes. Better to provide a redundant support system instead.

- In similar structures: where four points of support are provided consider only two as being effective.

- Avoid unnecessary eccentricity in bearings. If possible put them directly below the girder being supported.

3.4   Compassvale School, Singapore, 1999.

[Thung (2000)]

This case concerns the collapse of the roof of an RC school building while under construction. The school building was a two-storey multi-purpose hall. The first-storey was to be used as a canteen; the second was to be for school assembly and in-door games. The building was 42 m long, 27 m wide and 18 m tall. The roof was pitched and covered with clay tiles supported on timber battens and rafters, which were in turn supported on 10 steel trusses spanning 27 m. A cross-section is shown in Figure 3.9.

Collapse:

On 15 June 1999 when the roof structure was almost complete, it collapsed onto the first storey. Seven workers were injured.

Investigation:

The roof trusses, consisting of CHS (circular hollow section) members, were fabricated off-site in two half-sections, each 13.5 m long and 800 mm deep. The two half-sections were joined on site by welding, to form the 27 m truss. The designer specified these welds as 6 mm thick fillet welds. Each truss was bolted to the supports using four chemical bolts on each side (Figure 3.10). The support columns were 230 mm x 700 mm and were provided with 4No. T16 bars.

The designer apparently incorrectly modelled the structure. There were three possible models:

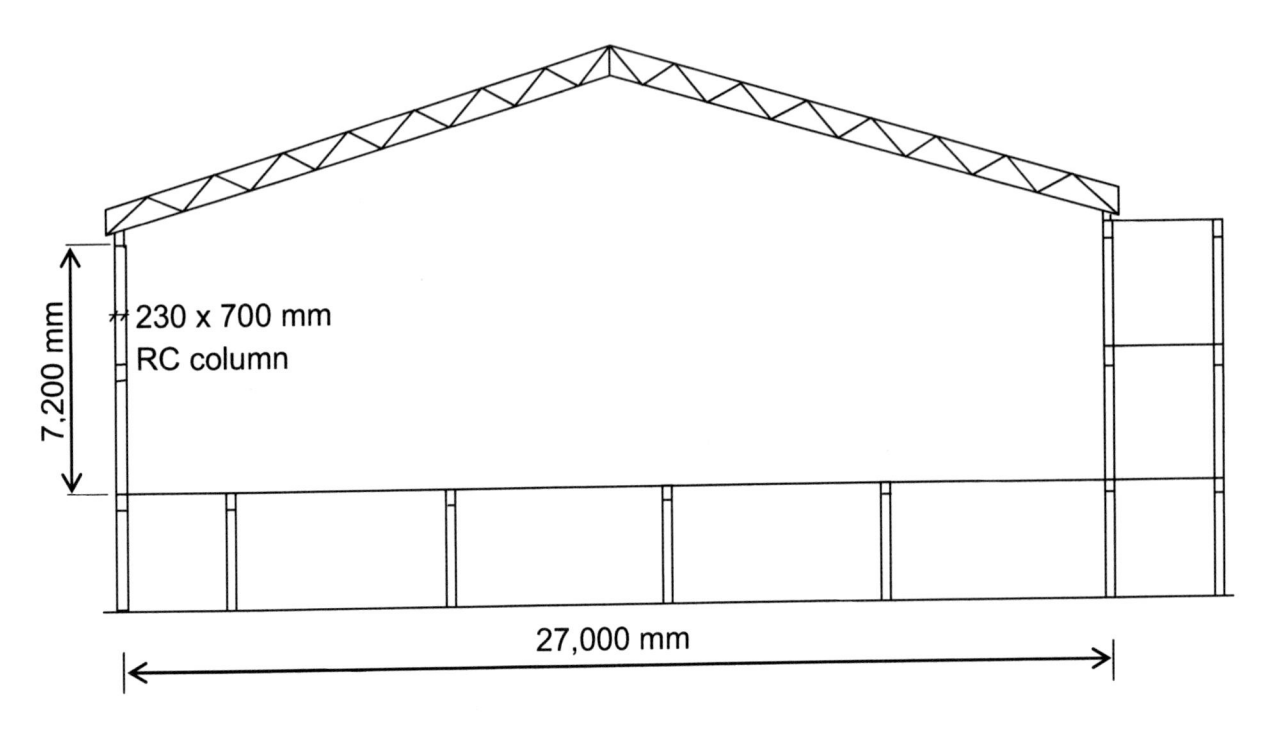

Figure 3.9: Cross section of school building

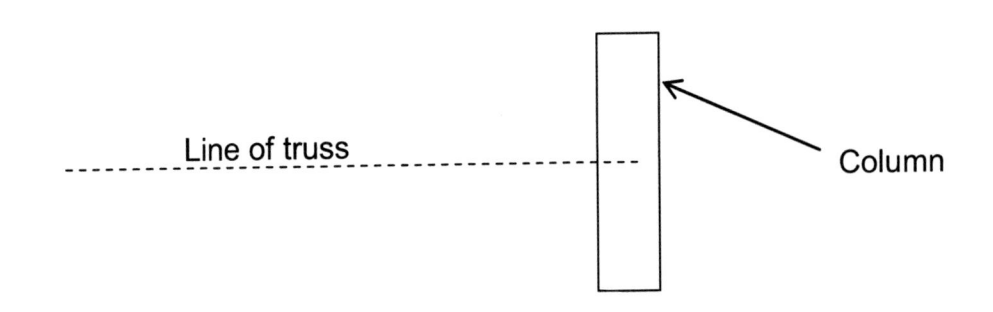

Figure 3.10: Plan of truss support

1. Simply supported truss: (Figure 3.11)

This required having a "pinned" end and a "roller" at the other end. Calculations showed that horizontal deflection at the roller was 163 mm under code specified service load. In addition the joint between the two half-trusses should have been designed for the full static moment. Clearly the details did not allow the roller-end to slide and the static moment was not designed for.

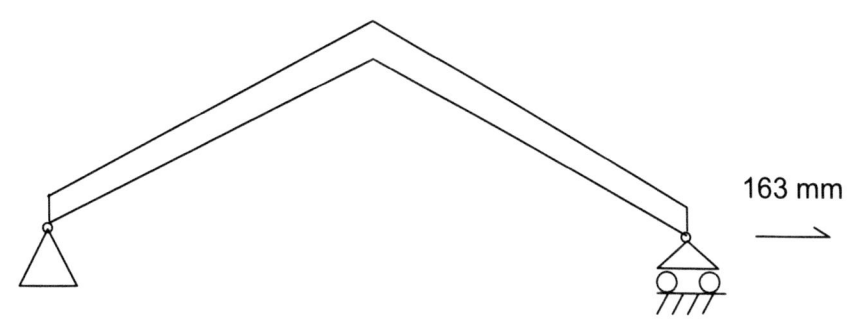

Figure 3.11: Simply supported truss

2. Trusses pinned to supports: (Figure 3.12).

The truss should have been designed to resist the resulting bending moments and axial compressions. The RC columns should have been sized to resist the resulting bending moments, etc. A movement of 147mm should have been expected and the finishes detailed accordingly.

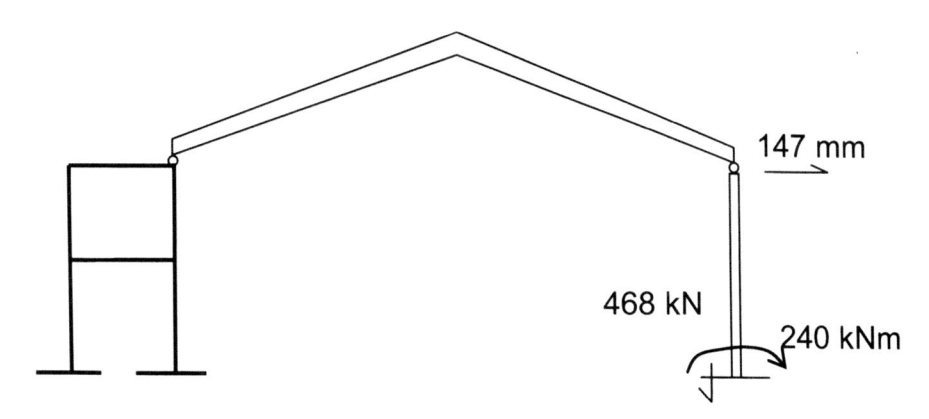

Figure 3.12: Truss pinned to supports

The connection between the trusses as well as the RC column itself would have been overstressed.

3. Three dimensional action of whole building: (Figure 3.13)

Analyzing this model gave similar answers to the two-dimensional model. In fact, about 10% more bending moment had to be resisted by the RC column. Again analysis showed that the RC columns would fail and roof truss splice welds would fail.

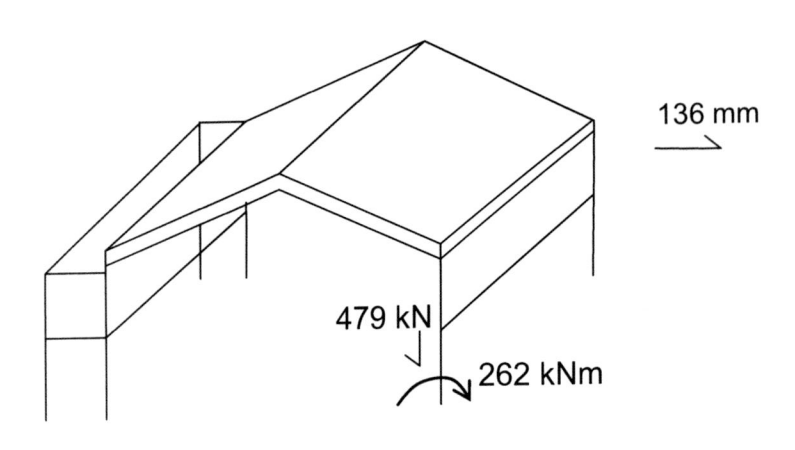

Figure 3.13: Allowing for whole building action

Unheeded warnings

The contractor reported to the engineer that in March 1999 that the roof beam at gridline along right-hand support had bowed out by 50 mm to 70 mm. He also reported that in May 1999 that the masonry walls along this gridline were out of plumb by as much as 130 mm; cracks were seen on the walls and RC columns.

Other errors

During assembly, it was found that the two half-sections of the roof truss did not meet properly. Circular steel rods were used in some locations to "fill the gap" before welding. The resulting welds were inferior. Supervision of site welding was inadequate, i.e., pre-welding inspections were not carried out and insufficient post-welding inspections were carried out.

Structural Engineering Failures: lessons for design

(4) In the designers' calculations it could be seen that simply supported behaviour had been assumed for the design of the members of the truss. The roof truss was designed to take dead load only (live loads and wind loads were neglected). The designers apparently designed roof truss separately from the supports and ignored their interaction. The checker (an engineer from another consulting engineering firm) did not check the calculations thoroughly and so did not spot these or any other mistakes.

Lessons

- Do not ignore warnings of collapse!

- The structure should be detailed to allow the assumed mode of behaviour to take place.

- To allow fabricated steel placed on *in situ* concrete to be adjusted to fit, allow movement joints (temporary, e.g., slotted-holes, if necessary).

- Pre-inspect areas to be welded and post-inspect welds properly.

- Roofs without a tie at eaves level are particularly prone to outward movements. [*Part of a roof which overhangs*]

- The top apex connection consisted of several CHS members meeting. This was a very difficult detail to construct as the members were circular and met at an oblique angle. The fit of the members must be very precise for welding.

- Over-simplification of design (the ignoring of any interaction of elements) can be dangerous.

- Calculations should be thoroughly checked but this checking should not be relied on to discover mistakes.

- Rely instead on **redundancy**. — *duplication of critical components increasing reliability*

Aftermath

Both Professional Engineers (one from the designer and one from the checker) were fined ($25,000 and $50,000 respectively) and lost their Professional Engineering licenses.

Neither was imprisoned. The judge said that he was lenient as nobody had been killed, (although one of the injured was permanently disabled).

"The checkers responsibilities are as great or greater than those of the originator"

Peter Campbell (former president of IStructE)

Chapter 4: Wind

"Everytime history repeats itself the price goes up"

Anonymous

This chapter consists of failures and near-failures where wind load was a problem. Five case studies are presented:

- Tay Bridge, Scotland, 1879;
- Tacoma Narrows, Washington, 1940;
- Ferrybridge Cooling Towers, England, 1962;
- John Hancock Tower, Boston, 1972;
- Citicorp Centre, New York, 1978.

4.1 Tay Bridge, Scotland, 1879.

[Ross (1984), Francis (1989), Petroski (1994); Petroski (1995), Shepherd and Frost (1995), Feld and Carper (1997), Kletz (2001), Jennings (2004), MacLeod (2005)]

The single-line railway bridge was over 1500 m from end to end. It consisted of 85 spans of malleable iron. Thirteen spans were through-trusses of up to 80 m span placed high above water (27 m) for navigation clearance, and were known as "high girders". The Firth of Tay, an inlet of the North Sea, is a bottle-shaped estuary. The bridge was at the neck of the "bottle". The topography of the land concentrated the northeasterly winds.

The bridge was designed by Sir Thomas Bouch, the leading bridge engineer of the day. He was under considerable pressure from the Client to save time and cost. Bouch consulted experts on what wind pressure to use for the design; they suggested various figures. He took the **lowest** of the suggestions. The bridge was designed for wind pressure of 0.5 kN/m^2 (For similar structures, contemporary American and French engineers were using a design pressure of about 2.5 kN/m^2). A section through a pier supporting a "high-girder" is shown in Figure 3.1.

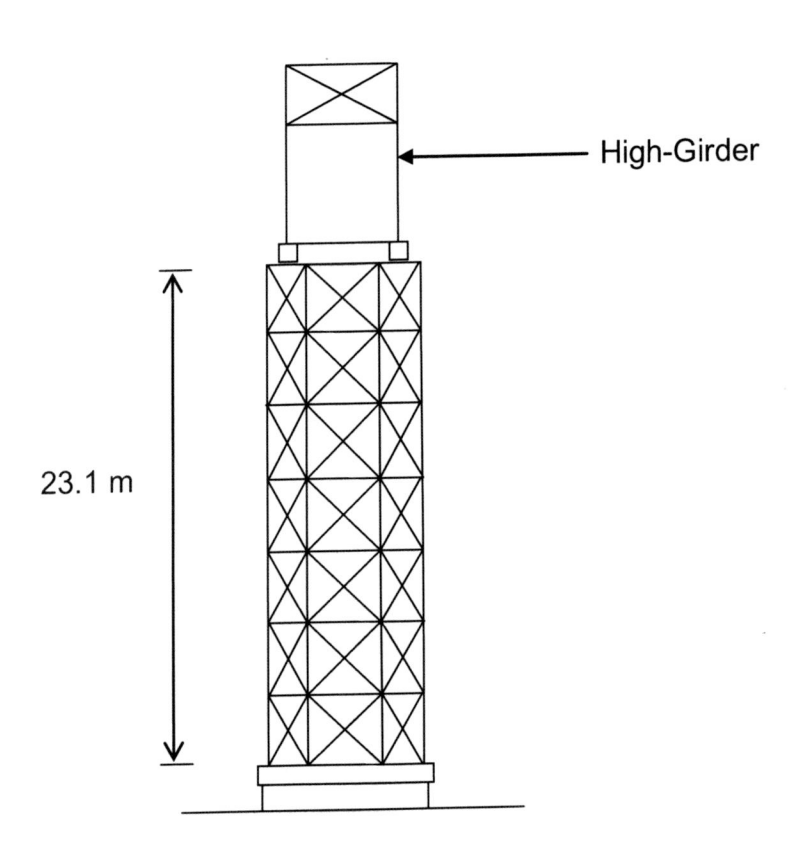

Figure 4.1: Section through pier supporting "High-Girder"

Collapse

Six months after opening: there were strong winds one winter night. Eye-witnesses said it was the worst storm in memory. The "high-girders" (13 spans) of the bridge collapsed as a train was crossing, killing 75.

Investigation

It was intended to construct the piers entirely of brick and concrete. During construction, however, surveys showed that the river bed was weaker than expected (due to misinterpreted boreholes). To lessen the weight (and save cost) cast iron pipe columns with cross bracing were used above the water. The casting and erection of the cast iron was not adequately supervised (supervision was by a very competent and honest **bricklayer** who had no experience in iron-work).

The quality of iron was generally poor with the thickness being uneven; blow-holes were present, and the lugs for the bracing were not secure. A lawyer wrote: "No man of skill apparently left to look after the ironwork of the bridge". Sir Thomas Bouch blamed entirely by the subsequent enquiry. The enquiry considered that the design wind pressure should have been 2.8 kN/m^2. British structures had to be designed for this 2.8 kN/m^2 until World War 1 (beginning 1914). This was criticized for being excessively conservative by engineers (e.g., Owen Williams). Later wind pressures required were reduced considerably.

Apparently the designer of this bridge had gone too far in the pursuit of economy in design and construction.

Lessons

- A wind tunnel test is the only way to determine likely wind pressure on a structure such as this as account of topography is taken.

- Do not allow commercial pressures to erode engineering judgment.

Aftermath

A new wider bridge (with two rails instead of one) was constructed adjacent to the old. The remaining spans were reused.

4.2 Tacoma Narrows Bridge, Washington, USA, 1940.

[Ross (1984), Dutt and George (1989), Francis (1989), Mark (1990), Levy and Salvadori (1992), Petroski (1992), Petroski (1994), Petroski (1995), Feld and Carper (1997), Wearne (2000), Billington (2006), Delatte (2008)]

Tacoma Narrows suspension bridge was opened in July 1940. It had a main span of 853 m and was 11.9 m wide (having only one traffic lane each way). Instead of the more conventional deep stiffening truss to distribute point loads, it had a 2.44 m deep I-beam supporting the concrete deck on each side of the roadway (Figure 3.2).

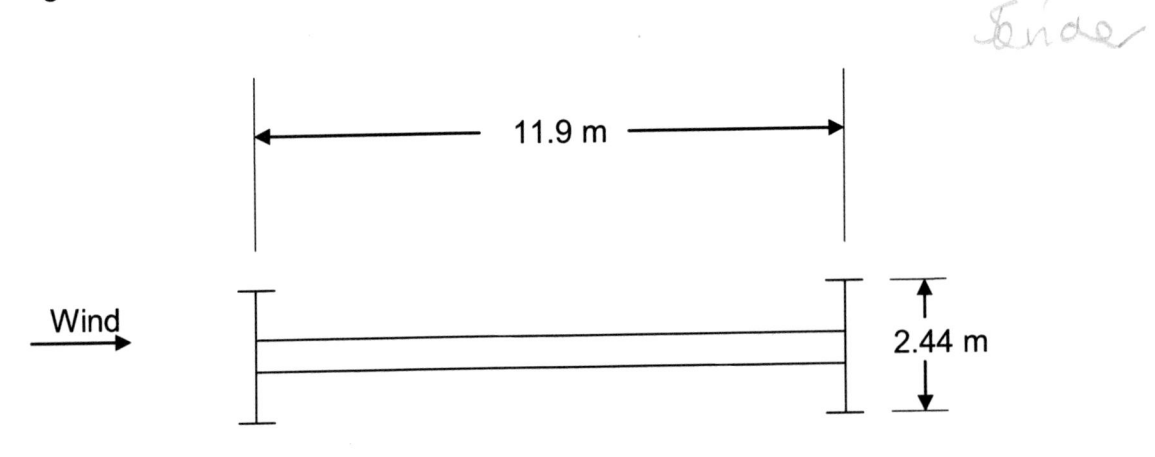

Figure 4.2: Section through deck

The designer was the world-renowned suspension bridge designer, Leon Moisseiff. It was designed using then-new "deflection theory". However, this theory considers only static behaviour. It suggested that vertical stiffness was the main consideration; that stiffening of the deck was not necessary if the cable system was sufficiently stiff vertically. It failed to note that wind requires the deck to have torsional stiffness too. A cheaper bridge resulted with a sleeker profile.

In early twentieth century the use of deflection theory was promoted by, e.g., Leon Moisseiff and Othmar Amman, so that conventional trusses stiffening deck could be avoided. If the stiffening truss were eliminated bridges could be sleeker (and cheaper!).

Thus at Tacoma, wind was considered as a purely horizontal static load. K-bracing was placed in the plane of deck to resist static effect of wind as horizontal load. However from the start it was clear that dynamic behaviour was important…Large vertical oscillations, even in light wind, earned the bridge the nickname "Galloping Gertie". These motions were not proportional to the wind speed.

Collapse

In November 1940, a steady 42 mph (19 m/s) wind blew. The motions quickly grew. Vertical vibrations up to 3.5 m in amplitude were recorded. However, soon twisting vibrations also began, twisting the deck to almost 45 degrees. The bridge twisted itself apart in less than an hour. The observed deformation was about 0.2 Hz. The only casualty was a dog in a car.

Investigation

An investigation committee was formed. Suspension bridge designer Othmar Amman (e.g., George Washington Bridge, and Bronx-Whitestone Bridge) and aerodynamic expert Theodore von Karman were members.

The committee found that the bridge designed for the **static** effect of wind only. The dead load was only 1/10 of that of any other major suspension bridge yet constructed. Tacoma Bridge was weak in torsion; it was more slender vertically (depth/span=1/350) and horizontally (width/span=1/71) than ever before. The bridge had no stiffening truss; ensuring it had 60 times less damping than a typical suspension bridge.

Aerodynamic instability of the deck was blamed for collapse. The movement of the bridge in response to wind forces increased the forces due to wind (a phenomenon similar to "flutter" of airplane wings). Aircraft designers knew wings of aircraft need to be closed sections so that they are very stiff torsionally; otherwise they are prone to torsional instability.

Thus the collapse of the bridge happened as a result of exceptional weakness of deck structure **in torsion**. Once the wind had an upward component the deck had little resistance to offer.

Experiments on models of bridge showed the squat H cross-section was especially vulnerable to flutter.

Torsional stiffness depends on:

- twisting of I-sections;
- bending of I-sections;
- distance apart of I-sections.

When a non-streamlined body is placed in an airflow, a series of vortices are released at regular intervals alternately from each side. These are known as "Von Karman vortexes". The release of each vortex imposes a small transverse force on the body. For Tacoma Narrows this meant the bridge was subject to a periodic force of about 1Hz. As the deck moved in response to these regular pulses, the bridge shed another set of vortices which happened to be of same frequency as torsional natural frequency of bridge (about 0.2 Hz).

Model tests later showed that the behaviour was very sensitive to the wind speed:

- At 38 mph: vertical movements were moderate (about a metre) with the two suspension cables moving in phase.

- At 42 mph: the period of oscillation suddenly increased threefold and the vertical movements were about 10 m. In addition, the cables moved out of phase with which the torsionally flexible deck could not cope.

The investigation committee concluded that *"if errors or failure occur, we must accept them as a price for human progress"*.

However, in the opinion of many engineers, it is unlikely this failure was due to the professions lack of knowledge. It was more a failure to apply knowledge we had already gained.

There were in fact many historical precedents for suspension bridges failing in wind (e.g., Menai Straits Bridge 1826; Brighton Chain Pier 1834; Wheeling Bridge in Ohio 1854; Niagara-Clifton Bridge 1889). Here, for example, is a quote from an eye-witness to the Wheeling Bridge collapse:

"Lunging like a ship in a storm, the deck rose to nearly the height of the towers, then fell, and twisted and writhed, and was dashed almost bottom upward"

One previous engineer who knew how to construct a suspension bridge to resist the wind (and also to support a railway) was **John Roebling (1806-1869)**. He designed the Niagara Gorge Bridge (1855), Cincinnati Bridge (1865), and the Brooklyn Bridge (1883). He wrote, in 1855, that stiffness of suspension bridge is achieved using "Weight, Girders, Trusses, and Stays".

Lessons

- Study of past suspension bridges would have revealed that wind was one of the main modes of failure.

- Wind is a dynamic force not a static one (gust period is typically 3 seconds and if the period of the structure is more, the wind should be considered a dynamic force).

- Frequently wind forces are neither horizontal only nor proportional to wind speed.

- Some shapes are unstable in the wind, resulting in larger forces when the shape moves in response to the wind.

- Structures of low natural frequency are most susceptible to the dynamic effect of the load.

- Do not ignore related fields, e.g., aerodynamics, which was seen as irrelevant to bridges, but was being used for aircraft development since 1930s.

- Designers were extrapolating from "successes" of the recent past (there had been no failures since 1889). "Absence of failure" is not same as "success". A latent failure mode may be triggered by yet-unexperienced conditions.

Aftermath

Tacoma Narrows Bridge was rebuilt using the original foundations. The new bridge is wider: it has four lanes, a stiffening trusses 7.5 m deep and dampers to stop oscillations. In addition the stiffening trusses are connected at the bottom thus completing the torsion "box".

Additional Information

Modern suspension bridges are built using decks of high torsional stiffness (i.e., greater torsional stiffness to ensure the natural frequency for torsion is high) and usually are wind-tunnel tested. Many are streamlined to reduce wind forces (European approach) or have deep open-web trusses instead of plate girders (American approach).

"Deflection theory" was later modified by Amman who used it to design later successful suspension bridges, e.g., Verrazanno-Narrows Bridge, New York. Bridges built before Tacoma using the unmodified theory were stiffened by replacing the girders with large stiffening trusses, e.g. Bronx-Whitestone bridge.

Lessons for other types of structure: e.g., canopy (Figure 3.3).

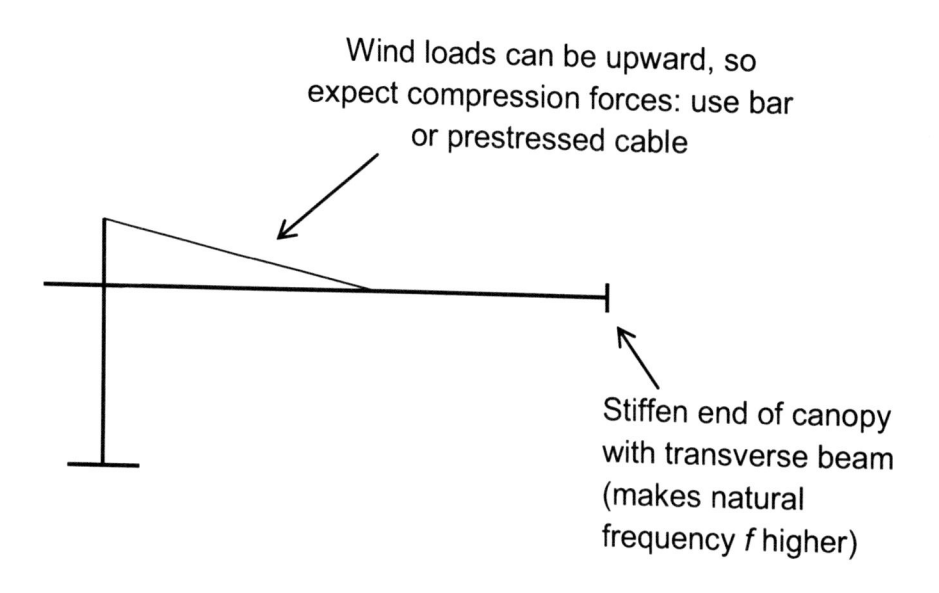

Figure 4.3: Canopy

4.3 Ferrybridge Cooling Towers, Pontefract, UK, 1965.

[Blockley (1980), Schlager (1994), Simu and Scanlan (1996)]

Eight "natural draft" hyperbolic cooling towers, each 114 m tall, a maximum of 88 m in diameter, and about 125 mm thick, were built at a coal-fired power station. They were built as thin shells of concrete: (an egg shell of the same diameter would be over 600 mm thick.) See Figure 3.4. All were constructed using a "Design and Build" contract in 1962. They had the largest shell diameter to date. They were reinforced using one layer of reinforcement placed centrally.

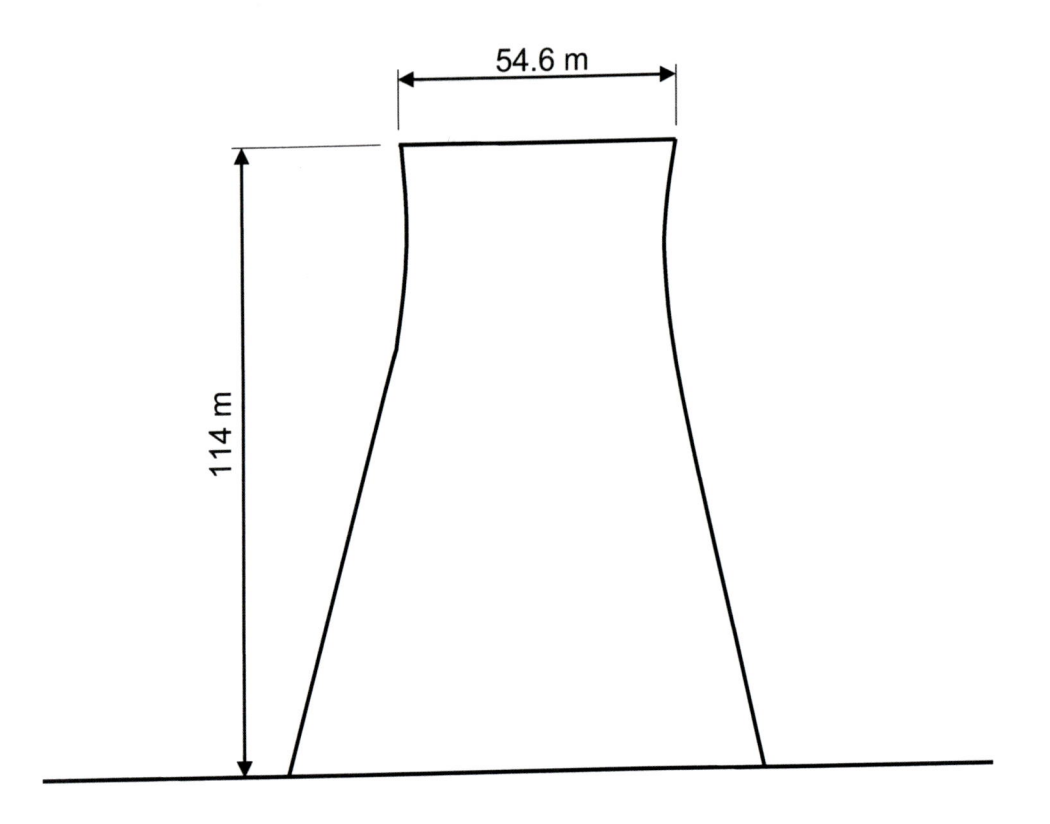

Figure 4.4: Elevation of one of the 8 towers

The spacing of the towers was closer than usual because of the need for support from coal seams below. The main load to be supported was self-weight and wind load. The specifications for the design indicated a wind speed of 28 m/s at 12 m height, based on engineering consultant's interpretation of wind tunnel testing report. No code of practice used to estimate wind.

Collapse

The shells were structurally complete but not yet in operation. A strong <u>westerly</u> wind (i.e., from the west) blew with gusts estimated later to be about 36 m/s (80 mph) at 10 m above the ground and 45 m/s (99 mph) at the top of the towers. This corresponded to a wind of 5-year return period for the area. The construction workers took a tea break.

The wind made the towers vibrate and produced a high-pitched whine "like someone rubbing a finger around the rim of a glass" in the words of an eyewitness.

In the space of an hour three shells on <u>east</u> side collapsed (Figure 3.5). Most of the remaining towers suffered large horizontal cracks especially near the base.

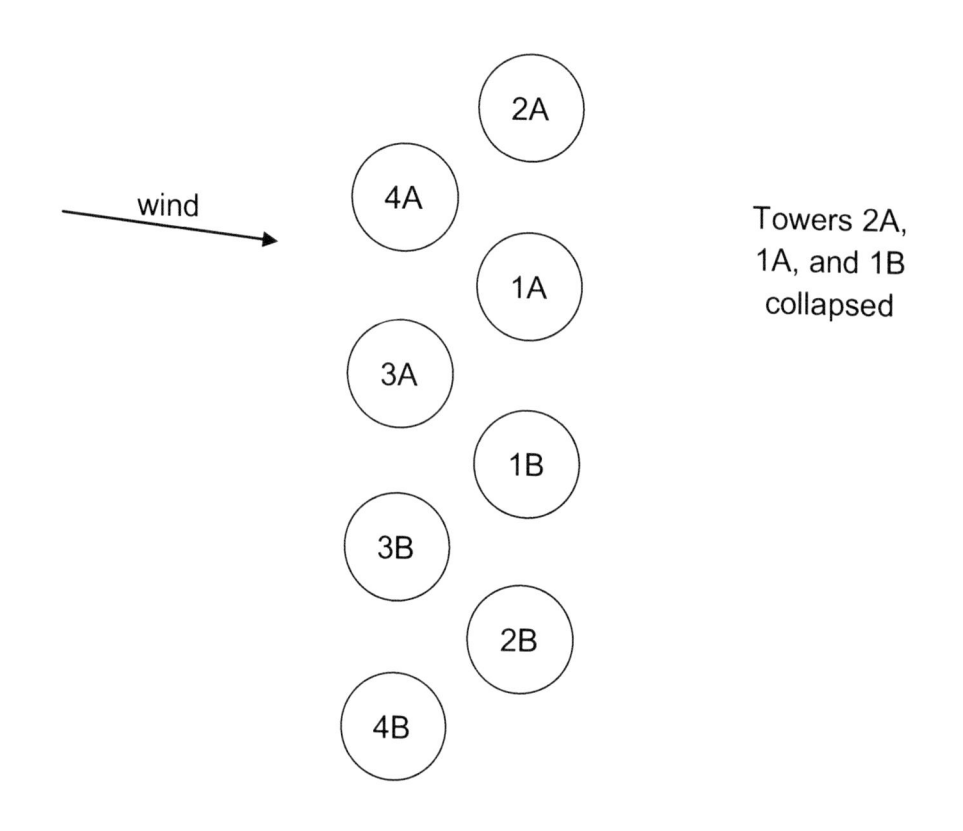

Towers 2A, 1A, and 1B collapsed

Figure 4.5: Plan of towers

Investigation:

The calculations were found to be numerically correct and the reinforcement was provided correctly. The construction procedure was also found to have been good.

However, specifications for wind loading were based on misinterpreted wind tunnel studies resulting in gusting being neglected and were conducted on an *isolated* tower. The effect of the closely spaced towers was to funnel wind onto leeward towers. This wind was both faster ("venturi effect") and more turbulent ("wake buffeting"). It excited resonance in the leeward towers. The amplitude of resonant vibrations grew over time until collapse occurred.

The specifications required design for "28 m/s at 12 m height". They did not specify the period over which this was to be average speed. In fact it was average over about 10 min. The shorter the averaging period the higher the speed (as the gusts are part of the average). If gusts are taken into account the correct speed for design is more than 50% more than the 10 min average. Thus the loads (which are proportional to the square of the speed) are at least 225% more! (Note: A more modern code, the British wind code CP3:1972, gives a 3-second gust speed for this area as 45 m/s at 10 m height with return period 50 yrs).

Thus the tensile stresses at the base of towers were greatly underestimated. It was these tensions that caused failure. In addition, the situation was made worse by the code used to actually design the tower (code in use for the RC design was British code CP114 a *working stress* code).

The following brief example will illustrate this disadvantage of the working stress approach: First, suppose the uplift due to wind was 100 kN and the self-weight was 90 kN (working values) then net uplift was 100 - 90 = 10 kN. According to the working stress codes, the safety factor for design was applied to **this** figure. Thus using a SF of **2** an uplift of 20 kN was designed for (working value). However, suppose now that the wind uplift was underestimated by 20% (so its actual value was 120 kN). Therefore actual net uplift was 120 – 90 = 30 kN. Thus failure would occur (as 20 kN allowed). Thus the design is very sensitive to changes in the value of major parameters. Using EC2 (a limit state code) and the load case 1.5xWL+0.9xDL the ultimate provision would be for an uplift of 99 kN.

Lessons

- Wind-induced stresses are considerably more severe in the case of groups of towers than for isolated structures.

- Tall structures typically have low natural frequency and so are very sensitive to gusts.

- Wake buffeting can only be safely ignored if the separation between structures is about 25 times their breadth or when their natural frequency is much higher than 1 Hz.

- Basing the design on the difference between two large numbers is risky if one of the numbers could be in error. Conduct sensitivity analysis to see how big the errors in design would be if the loading data is changed (this could happen in any case if a new structure was built in the neighbourhood, increasing turbulence in the "shed" or "second-hand" wind).

Aftermath

A second layer of reinforced concrete was added to the outside of the remaining shells, doubling the thickness. The collapsed shells were re-built with double the thickness. Personnel are not allowed to go near to a cooling tower shell in a wind storm.

Much research on wind loading resulted from this failure.

4.4 John Hancock Tower, Boston, USA, 1972.

[Ross (1984), Mark (1990), Levy and Salvadori (1992), Cambell (1995), Petroski (1996), Simu and Scanlan (1996), Feld and Carper (1997), Taranath (1998),]

All early high-rise buildings had masonry external walls and partitions, so were heavy and stiff. Thus the accurate determination of wind effects was of little importance. Modern high-rises are lighter and less stiff. Wind effects are thus of more importance.

The John Hancock Tower is a 60 storey steel (234 m) building completed in 1972. The plan is constant over the height of the building and is shown in Figure 3.6.

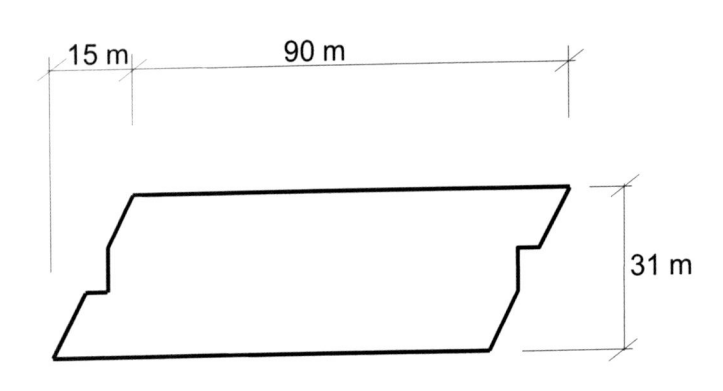

Figure 4.6: Plan of John Hancock Tower

The design satisfied all code requirements. The steel frame was proportioned such that the top deflection was equal to building height/400 in each direction under 50-year wind loads. Reflective double glazed glass completely covered the tower. However, even before construction was completed glass started to fall. Delays and design modifications doubled the original construction cost.

New wind-tunnel tests confirmed wind was not the reason for the glass failure, and also showed that upper floors would experience a high acceleration both in the <u>short direction</u> of the building and in <u>twisting</u> (as bending and torsional natural frequencies were similar). To counter this two dampers were installed near the top of the building. Each was a 3000 kN block of lead resting on a film of oil and attached to the structure using springs. These dampers dealt with the <u>comfort</u> problem.

A world expert on building dynamics, was consulted regarding the <u>safety</u> of the tower. He identified a P-Delta problem in the <u>long</u> direction of the building; the additional moment due to second order effects, not recognized in design, was nearly 50% of the first order moment (M_1), making the Modification Factor (*MF*) nearly 1.5 (Figure 3.7).

So 1,650 tons of new steel diagonal braces were added to the core, in order to *double* the stiffness in the long direction of the frame.

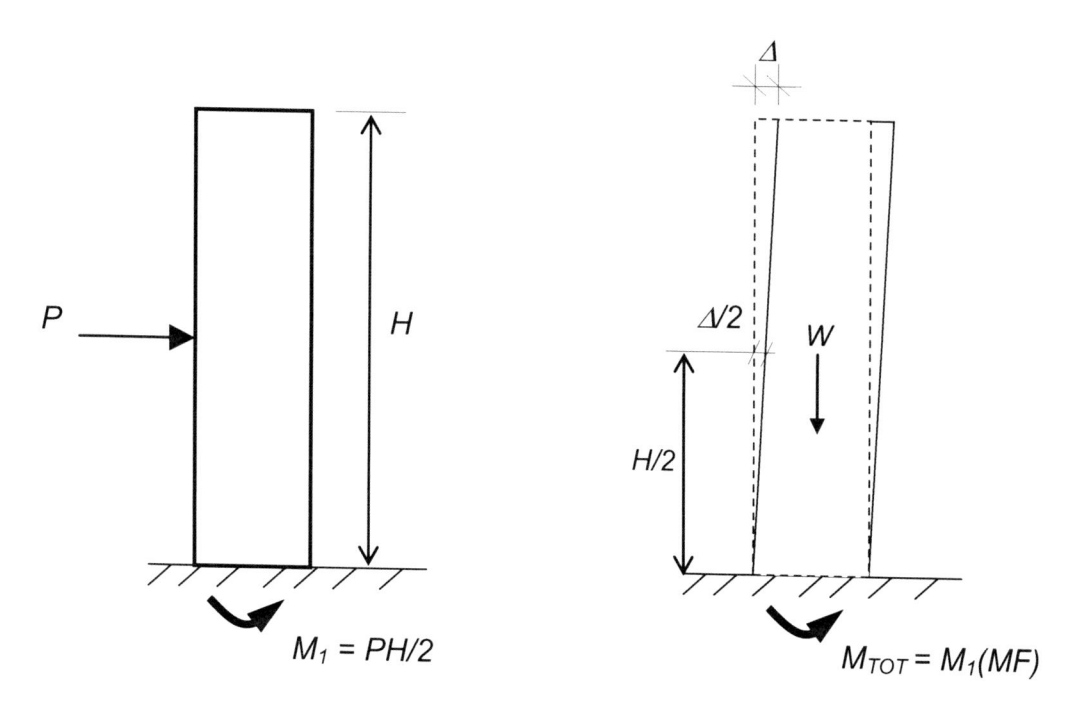

Figure 4.7: P-Delta phenomenon: effect of lateral load magnified by deflection

However, the glass continued to break. Eventually the architect's office discovered why. Thousands of cycles of heating and cooling had caused a fatigue crack in the solder of the lead spacer which separated the two panes of glass, to migrate into the glass. So all of the 10,344 panels were replaced panels of single thickness tempered glass.

Lessons

- Avoid multi-storey structures that are too flexible laterally; the P-Delta check is very important.

- Long buildings and/or unusual plan shapes can cause unexpected wind effects (e.g., torsion).

- Do not rely merely on the use of a code. The code is the *minimum* acceptable standard for *conventional* structures.

- High-rise buildings of more than 25-30 storeys should be tested in a wind tunnel especially if the slenderness is high (> 5).

- These tests often pay for themselves in the long run as the overall forces may be less than the code values.

- A damper should not be relied upon to attempt to solve a safety problem. This is why they solved the P-Delta problem by adding stiffness.

- Avoid having similar natural frequencies in bending and torsion.

Additional Information

The Boundary Layer Wind Tunnel at the University of Western Ontario was set up in 1965 by Alan Davenport. The first high-rise structures to benefit from accurate wind tests in their design were the World Trade Centre (opened 1973) and the Sears Tower (opened 1974).

The P-delta effect is the reason that many engineers ensure that all concrete structures, particularly those with flat plate/slab floors (i.e., especially heavy construction), are braced with shear walls.

The Eurocode for concrete, EC2 (108), gives a method to ensure sufficient shear walls so that the columns can be considered as braced. The effect of this is that the modification factor *MF* is about 1.03.

According to Robertson (1987), an estimate of the Modification Factor (*MF*) is given by the formula:

MF = 1/(1-*WR*/*Q*))

where "*W*" is building weight (in Newtons); "*R*" is drift ratio (e.g., 1/500); and "*Q*" is total wind load (in Newtons). Well designed multi-storey buildings should have an *MF* no more than about 1.1 (i.e., the total moment should not exceed the first order moment by more than about 10%).

Even buildings with symmetrical layouts can experience torsional problems, especially if the plan dimensions are large in which case the wind pressure can be eccentric to the centre of resistance.

"Hancock...was comparable to Tacoma Narrows...it really changed that whole aspect of building engineering. It was a symbol."

Paul Weidlinger, founder of "Weidlinger Associates" a firm of consulting engineers.

"Any time you depart from established practice, make ten times the effort, ten times the investigations. Especially on a very large-scale project"

William LeMessurier (consultant for design of dampers) after Hancock.

Similar cases of multistory buildings experiencing problems with the wind:

Case 1:

- In 1983 a 47-storey steel building in Houston, Texas (designed to ACI-318) suffered a near collapse, caused by P-delta effects, during Hurricane Alicia (Chiles 2002).

- This building was built in 1971.

- In 1994 it was strengthened substantially (at a cost of about US$15 Million) to cure its excessive deflection (Colaco et al. 2000).

Case 2:

(Zallen 2004)

- In 1984 a 7-storey hotel building with steel columns and post-tensioned floors was constructed using the liftslab technique in Portland, Maine.

- The building had steel cross bracing designed to resist an average wind load of 0.5 kN/m^2.

- Construction using liftslab technique meant there was little resistance to lateral load from frame action.

- While still under construction, outside consultants calculated it was excessively flexible under this wind load.

- The P-delta effect meant the *MF* was 1.36.

- The bracing was stiffened so that the *MF* was reduced.

Case 3:

(Ratay 2010)

- A 40-storey steel building was constructed in 1967 at 28 State Street, Boston.
- The building has a rectangular plan.
- The owner could not get tenants for the building because of the high torsional accelerations on upper floors.
- The first three natural frequencies were similar.
- The building was retrofitted with 40 viscous dampers in 1995.
- The cost was US$600,000.

Case 4:

(Ratay 2010)

- An 11-storey steel building was constructed in the 1990s.
- The resistance to lateral loads is from braced frames;
- The centroid of resistance was eccentric resulting in large torsions;
- The torsional accelerations were unacceptable.

4.5 Citicorp Centre, New York, 1978.

[Chiles (2002), Simu and Scanlon (1996), Morgenstern (1997), Petroski (1996), Martin and Schinzinger (1997), Delatte (2008)]

Citicorp Centre was completed in 1977. It is a slender steel framed tower. (Slenderness Ratio = 6). It has a height 279 m (59-storeys), square plan on upper floors. A church had to be accommodated on the site (the original occupant of the site). This gave rise to the unusual plan: the building sits on nine-storey stilts placed at <u>middle</u> of the sides.

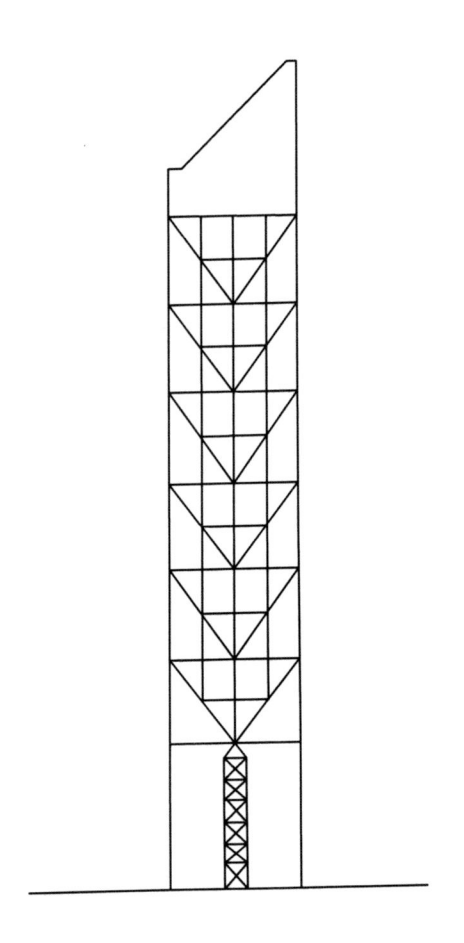

Figure 4.8: Citicorp Centre (now Citigroup Centre)

The tower's engineer was William LeMessurier well-known for his innovative approach. It was a very light-weight tower, having a density of only 130 kg/m^3. This compares with a typical value for a conventional steel building of about 190 kg/m^3. To reduce the accelerations that are caused by the wind on such a light structure, a tuned mass damper (TMD) was installed during construction at top of the tower (a 400-ton concrete block).

Gravity and wind loads were resisted by system of diagonals. These diagonals channeled the loads to the centre of each side. The design completely satisfied the code (New York Building Code) requirements (Figure 3.8).

The building was completed and occupied in 1977. In June 1978 a student asked LeMessurier by phone why the supports were placed at the middle of the sides rather than at each corner. LeMessurier explained about church and said the position supports makes the building especially resistant to "quartering winds", i.e. those along building's diagonal.

After the phone call LeMessurier decided to use the topic for a forthcoming structural engineering class he taught at MIT. So he worked out some numbers for the building resisting quartering winds. To his surprise, he found that far from the configuration being more efficient, the diagonal bracing was highly stressed under strong quartering winds.

When quartering winds were considered the members that had been in net compression went into tension. LeMessurier phoned one of his staff directly concerned with site operations. He found out that connections were bolted instead of welded (the original requirements of the design were that all connections were welded). This meant that some of the tension connections were under designed by 160%.

The dilemma faced by LeMessurier was that the building satisfied the code (which only required consideration of wind perpendicular to the face), but was actually potentially unsafe when quartering winds were considered. This was because the originally specified welding would have meant the full member strength was available, while the bolts just catered for the member force previously calculated ignoring quartering.

In July 1978 LeMessurier ordered new wind-tunnel tests to examine the problem further. A boundary layer wind tunnel was used. Tests were run on the building with the tuned mass damper on and off. It was found that *if* the TMD was not functioning the building had a 50% chance of blowing down (connection failing due to overstress) in a sustained wind of 70 mph. This is a 16-year wind for New York. *If* the TMD was functioning the critical wind had a return period of 55 years. The winter was approaching, so LeMessurier knew he had to act fast.

LeMessurier was prepared that he would face litigation, bankruptcy, and professional disgrace. Nevertheless, he blew the whistle on himself. He contacted the architect and his insurance companies lawyers. Leslie Robertson was appointed a consultant by the insurance company.

Citicorp did not spend time apportioning blame. The strategy adopted for the repair was to ensure operation of the TMD (by installing emergency generators) while new steel plates were welded over the connections. In case a large storm occurred before the repairs were completed, a plan was devised for evacuating Citicorp Centre and large area around it (the Red Cross estimated there might be 200,000 deaths otherwise).

A group of weather experts was hired to provide predictions of wind speeds. Technical staff from the TMD manufacturer were hired to attend to the machine around the clock in case it failed. Welding began in August. It was done at night in case the smoke and the smell would alarm the tenants and set off smoke detectors.

A plywood box was made for each connection so the welder could work inside without damaging the surrounding decorations. Tenants were told the reason for repairs was that new data indicated connections needed to be strengthened and that Citicorp were being extra careful.

Fortunately no storms hit the area during the repair period (about two months). When the repairs were completed the building could resist a seven-hundred-year storm event (at ultimate). The cost was roughly $8 million for the work. LeMessurier's insurer agreed to pay

$2 million. Citibank accepted this $2 million. No litigation ever followed. LeMessurier's reputation for honesty and competence was enhanced. The following year his insurance premium actually went down! After all, in the words of one of his staff, he "had prevented one of the worst insurance disasters of all time".

Lessons:

- Many codes require that buildings are designed to resist winds acting perpendicular to the building faces. In some circumstances the nature of the bracing means it is more susceptible to quartering winds.

- It is risky to design strictly to codes, which cover the "ordinary", when the building is far from ordinary.

- A Tuned Mass Damper should not be considered a safety device, i.e., its ability to reduce deflections should not be relied upon in a large storm. (E.g., after the 1993 bombing of the World Trade Centre in New York emergency generators lasted only 15 minutes before failing due to loss of cooling water.)

- Thus a TMD cannot be used to solve a **safety** problem.

"I have a lot of admiration for him, because he was very forthcoming. While we say that all engineers would behave as he did, I carry in my mind some skepticism about that"

(Robertson of LeMessurier)

"You have a social obligation. In return for getting a license and being regarded with respect, you're supposed to be self-sacrificing and look beyond the interests of yourself and your client to society as a whole. And the most wonderful part of my story is that when I did, nothing bad happened"

(LeMessurier)

Chapter 5: Maintenance

"Human history becomes more and more a race between education and catastrophe."

H.G. Wells (1866-1946) in "The Outline of History", Vol.2, Ch. 41 (1920).

This chapter contains case studies where the lack of maintenance played a role in the collapse.

The case studies are as follows:

- Point Pleasant Bridge ("Silver Bridge"), 1967.
- Mianus River Bridge, 1983.
- Antioch School, California, 1980.
- Ynys-y-Gwas Bridge, Wales, 1985.
- Schoharie Bridge, New York, 1987.
- Piper's Row Car Park, Wolverhampton, UK, 1997.
- De la Concorde Overpass, Montreal, Canada, 2006.

5.0 Introduction

[Francis (1989), Gordon (1988)]

Brittle Failure of Steel Structures:

Brittle fracture is an explosive type of failure that takes place due to crack growth. No yielding takes place. There are two stages:

1. Crack growth stage: slow growth of crack, possibly over many years;
2. Fracture stage: crack extends at about the speed of sound.

Consider a plate under uniform tension stress σ (Figure 5.1)

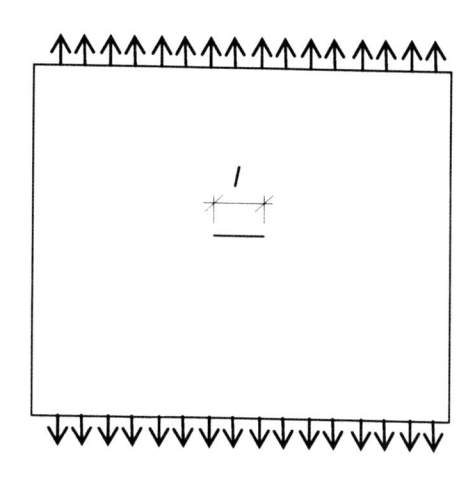

Figure 5.1: Plate with crack

Explosive, brittle fracture will occur once a crack reaches a "critical length". For a homogenous isotropic material the critical crack length (l_g) is roughly equal to

$l_g = 4 WE/\pi\sigma^2$

where E is Young's modulus, W is the "Work of Fracture", and σ is the average stress.

The lower the work of fracture, W of the material or the higher the stress, σ then the smaller is l_g. The work of fracture, W is measured in the Charpy impact test. A short length of the material is broken by the impact of a hammer that swings on the end of an arm. The amount of energy absorbed in fracture by the specimen is indirectly measured by the amount of up-swing of the machine's arm. Thus W is a measure of ductility; a higher figure indicates more ductility.

Typical values of W (measured in J/m^2) are as follows: Copper 2×10^5; Mild Steel (at 25°C) 10^5; Mild Steel (at −100°C) 10^4; High Carbon Steel 10^4; Cast Iron 10^3-10^4.

In mild steel structures ($W = 10^5$ J/m^2), stressed to about $\sigma = 165$ N/mm^2, cracks of about 1 m long are still shorter than critical length and so are <u>safe</u> as there is no risk of brittle fracture.

Thus the safe stress for design is the lower of:

- Yield Stress/F.S., where the factor of safety, F.S., is usually about 1.5, and,
- The stress that implies a crack length that can be readily detected by an inspection (usually visual), e.g., 0.5-1 m.

Note that the **larger** the structure then the **smaller** is the safe stress as it is physically possible to have longer cracks.

Cracks grow longer as a result of fatigue and/or corrosion. When steel structures are inspected, corrosion and cracks are being looked for. This shows the importance of maintenance of steel structures. Two important factors influencing steel ductility are temperature and the degree of triaxial tensile stress present. The effect of temperature is shown graphically in Figure 5.2.

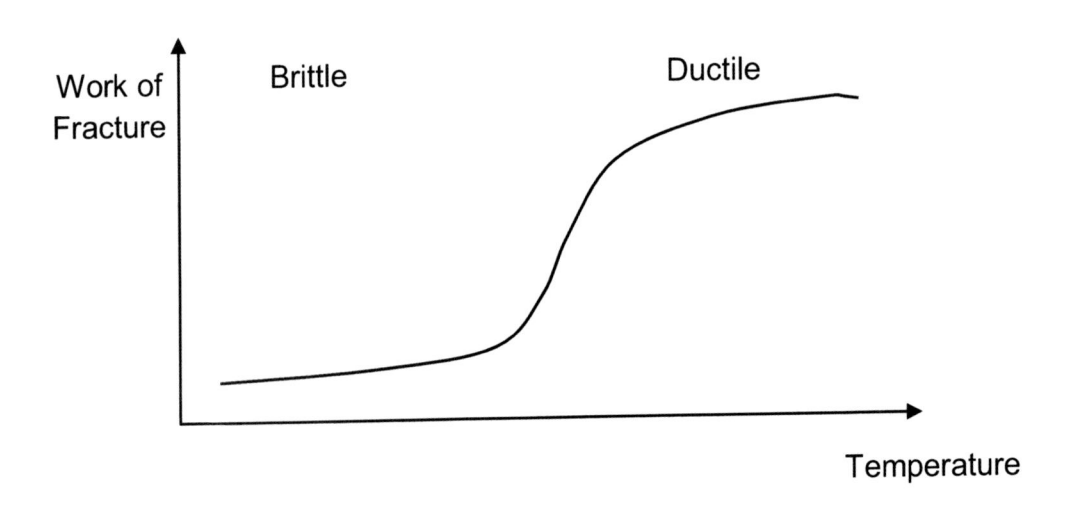

Figure 5.2: Effect of temperature

Effect of triaxial tension:

Steel is made more brittle if the tension is triaxial. This is because there is less shear to move the dislocations of the material and so ductility is reduced. The worst case is if the steel is under equal tensions in each direction. Then there is no shear and so the steel behaves as a brittle material. Discontinuities in thick plates (e.g., holes, steps in plates more than about 25 mm or so) make triaxial tension there more likely.

The following figure shows a <u>thick</u> plate under a uniform tension stress σ (Figure 5.3).

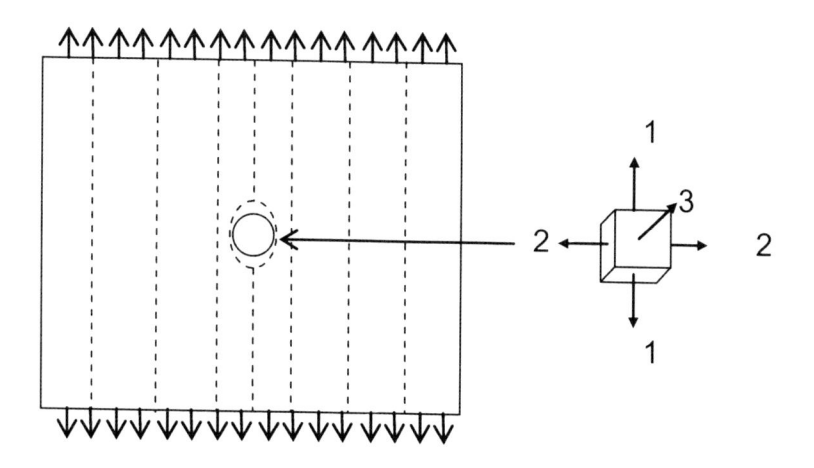

Figure 5.3: Triaxial stresses in a thick plate

The stress trajectories are shown as broken lines. The hole interrupts the trajectories as the tension must "go around" the hole. Direction 1-1 is not quite vertical. The tension in direction 1-1 requires for equilibrium one in the direction 2-2. In addition, the stress in 1-1 causes a contraction at 90^0 (i.e., 2-2 and 3-3) due to Poisson's ratio. In a thick plate there is significant resistance to this tendency to contract. Thus the restraint induces more tension in 3-3. So a small hole ensures that triaxial tension is experienced near the hole and the plate will not be ductile there. Thus the W is lower and so is l_g.

5.1 Point Pleasant Bridge, Ohio/West Virginia, USA, 1967.

[Silby & Walker (1977), Ross (1984), Fontana (1986), Levy & Salvadori (1992), Petroski (1992), Lichtenstein (1993), Schlager (1994), Shepherd & Frost (1995), Feld & Carper (1997), Wearne (2000), Delatte (2008)]

The bridge was built in 1929. It was a steel eye-bar suspension bridge. The main span was 214 m, and the side spans were 116 m. (See Figure 5.4). It was known as "silver bridge" because of its shiny aluminium paint. The "cables" of the bridge consisted of sets of two parallel eye-bars pinned together, like a bicycle chain.

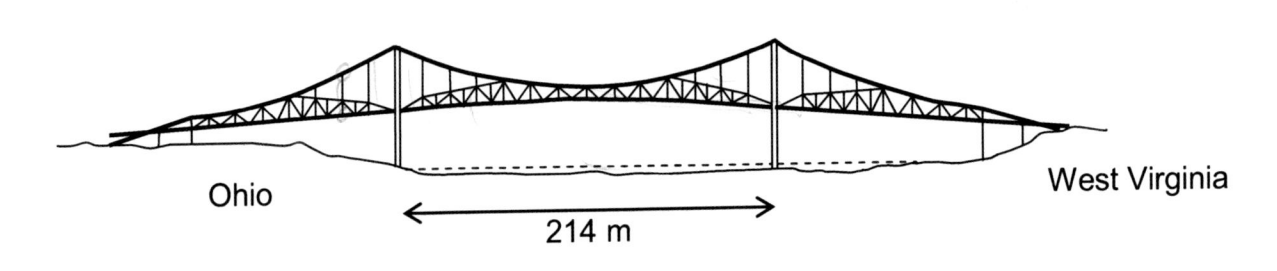

Figure 5.4: Point Pleasant Bridge

The eye-bars were 15 m long, 50 mm thick and 300 mm wide and made from high-tensile, heat-treated steel. It was an innovative design: for economy the suspension system was also used as the top chord of the stiffening truss. Truss members were directly attached to a pin (290 mm in diameter) at each connection. The pin was retained in place by circular covers and held by a bolt as shown in Figure 5.5. The bridge was unique in two respects: it was the first to use eye-bar chains and first to use high-strength, heat-treated carbon steel.

Collapse

The bridge collapsed suddenly in December 1967 during an evening rush-hour. The chains broke; one side span fell; the main span flipped over, dumping traffic into the river. Then the other side span and both towers also fell. There were 37 vehicles on the bridge when it collapsed. Forty-six people died and nine were injured.

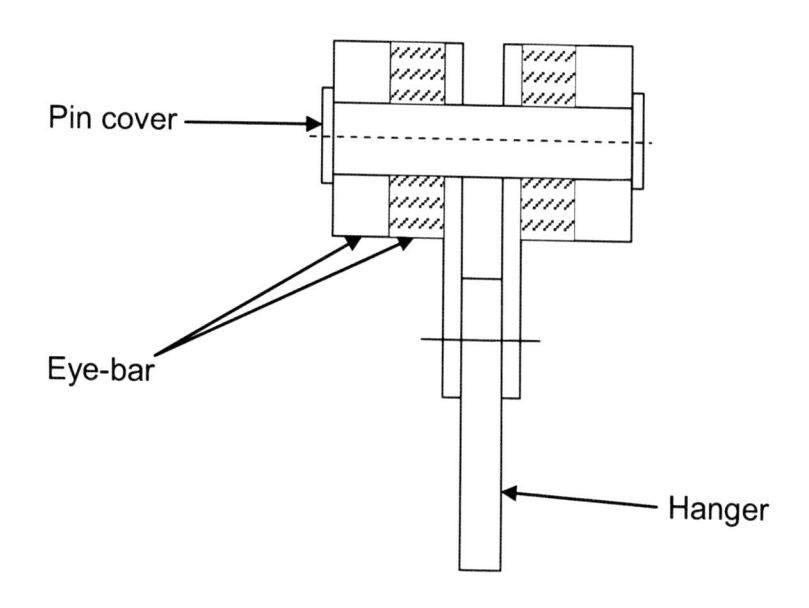

Figure 5.5: Pinned connection

Investigation

The original design was conservative. Thus the stresses were not excessive even for the live load the bridge was carrying when it collapsed. The theoretical stress concentration factor for stresses at the edge of a circular hole of three was taken into account in the calculations. Eventually the failed eye-bar was found. It had come from next to the Ohio-side tower. Inspection showed the failure initiated near the pin of this eye-bar. The companion eye-bar was heavily bent by the pin as it pulled out.

It was found that the fatal flaw initiated in the eye-bar probably during manufacture. The flaw was concealed by the pin-cover. Over the years, pollutants entered in the gap between the pin and the eye-bar (a gap of 3 mm) and corrosion took place. The crack grew bigger. Corrosion made the effects of stress and fatigue worse and eventually the critical crack length was reached and so one eye-bar failed.

The high-strength steel that the eye-bars were constructed from was supposed to have plenty of ductility. (Eye-bar steel yield = 517 MPa, working stress = 345 MPa (F.S. = 1.5)). However the heat-treating of large, thick sections is difficult. If not done properly, it can result in slow cooling of the interior and so a loss in toughness as the grain size is allowed to be large.

High ductility means the critical crack length is long. On the day of collapse temperatures were low (about -1^{0}C). This would have made the steel more brittle and so the critical crack length required to initiate brittle failure was lower. Thus the fatal crack could reach its critical length more easily. It was estimated that the critical crack length was only 6 mm.

The bridge was last inspected in 1965. It was a visual inspection using binoculars to view the bridge elements. The inspection personnel were not properly trained. No remedial work was recommended then. The last full inspection was in 1951. Eye-bars were made from the then newly developed heat-treated steel but were never tested. Instead small samples of similar steel were tested.

Thus the cause of the collapse was stress corrosion/corrosion fatigue, allowing growth of a flaw of critical size. Clearly, the inspection system failed. The details of the connection made it difficult to inspect properly without dismantling it. Even modern NDT techniques could not inspect such a connection.

Lessons

- Connections should be not be difficult to inspect or encourage corrosion (the connection trapped dirt behind the pin-cover and so encouraged corrosion).

- Bridge inspection must not be left to non-technical personnel.

- Bridge lacked redundancy: when single eye-bar failed whole bridge collapsed.

- Modern suspension bridge cables contain thousands of separate wires so are much more resistant to loss of one.

Additional Information

Systematic inspection of bridges throughout the USA began in 1968 as result of this failure. Every bridge was required to be inspected at least every two years.

As a precaution a similar bridge over the Ohio River (St Mary's) was demolished soon after this failure.

5.2   Mianus River Bridge, Connecticut, USA, 1983.

[Levy & Salvadori (1992), Petroski (1992), Schlager (1994), Shepherd & Frost (1995), Petroski (1995), Feld & Carper (1997), McEvily (2002), Delatte (2008)]

Mianus Bridge was a steel road bridge, skewed at 53° to the direction of travel. It was constructed in 1959. It consisted of parallel steel I-shaped plate girders continuous over supports. Every other span had a statically determinate suspended span linking the continuous spans. The suspended spans were pinned at one end and hung at the other end by means of a pin-and-hanger assembly.

Collapse

At about 1:30 am one morning in 1983, a 30 m long three lane wide section (an entire suspended span) suddenly fell out of the bridge. The 500 ton section that fell consisted of two I-girders and four cross beams. The timing was fortunate as the road was very busy during the day. However, three people were killed and three injured.

Maintenance

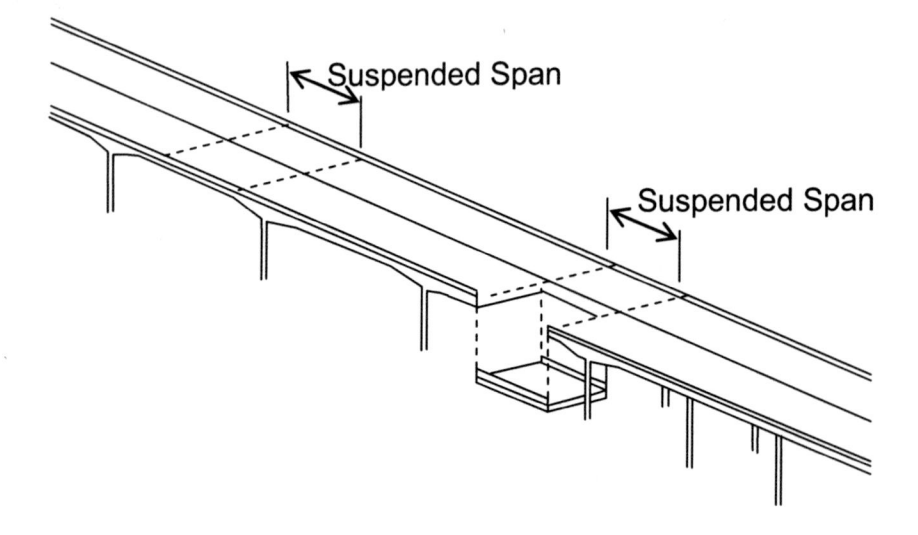

Figure 5.6: Mianus Bridge

Investigation

Inspection teams were required to inspect nearly 300 bridges per year in Connecticut. Unfortunately, the truck with the inspection boom needed to inspect Mianus River Bridge was out of commission, so the underside was inspected using binoculars from the river 22 m below.

It had been recognized that the determinate nature of the suspended span support system was potentially unsafe since 1968 (i.e., since the collapse of Point Pleasant Bridge). Slings had been added to many similar bridges constructed before then, but not to Mianus River Bridge.

The remains of the broken pin were found in the connection. The 180 mm dia. pin was held in place using a bolted-on cap plate. Many cap plates were dished outwards from the build-up of rust behind. A heavy concentration of rust was found throughout the hanger assembly.

The extensive rusting behind the cap plate would not have been visible to any inspector unless the connection was dismantled. The dishing would have been visible but it was apparently thought insignificant. It was found that around ten years before the collapse the drains had been paved over so water containing salts flowed down through the joint worsening the corrosion (see Figure 5.7).

Acceleration or braking of vehicles is rarely centered on bridge and so they impose a sideways force on the bridge, and thus tend to push the hanger plates off the pins. The skew of bridge increased these forces.

Thus, it appears that the rust caused the pin retaining bolt to fail in tension and the pin to slip from the hanger.

Lessons

- Do not make large structures statically determinate unless there are adequate precautions to improve post-collapse behaviour.

- Large structures should be thoroughly inspected by trained staff regularly.

- Avoid details which make inspection difficult and/or promote corrosion (e.g., by trapping dirt).

- Make sure the bridge is properly drained especially if salts are used to de-ice the deck.

Maintenance

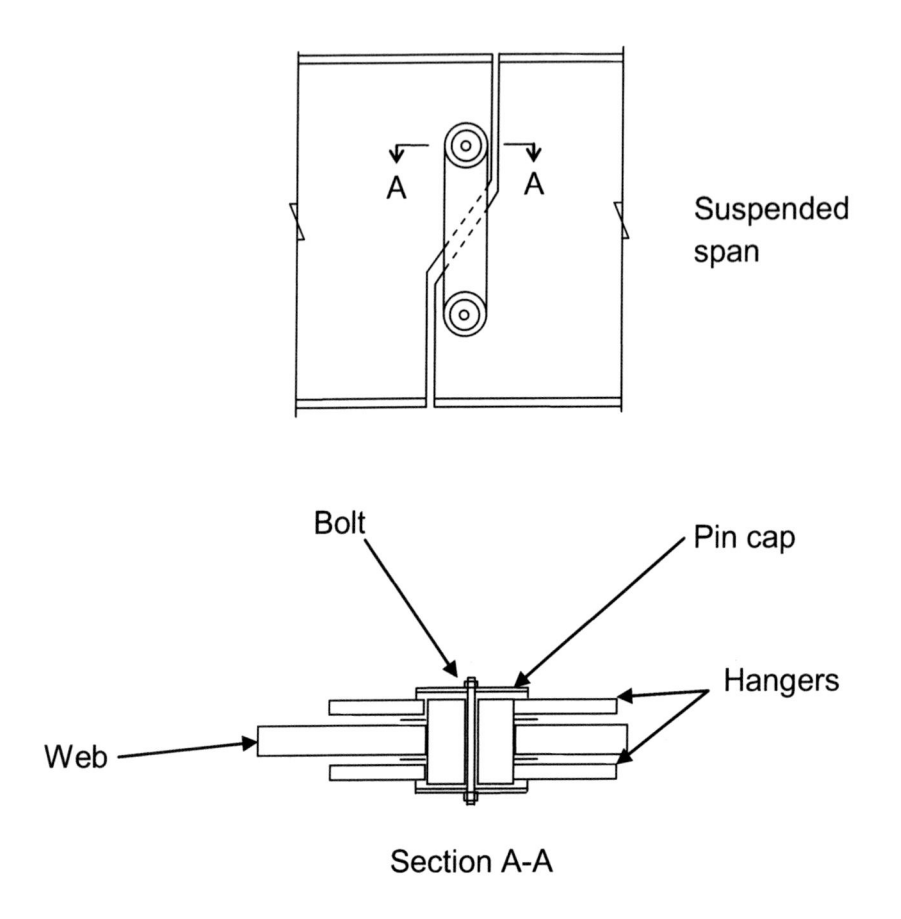

Figure 5.7: Pin and Hanger Assembly

Aftermath

A court found that deficient inspection was the main cause of the collapse. Connecticut DOT repaired hundreds of bridges in this collapses aftermath. The budget given was five times that before collapse.

Improving Corrosion Resistance by Good Detailing See Figure 5.8 (Dowling et al. 1988)

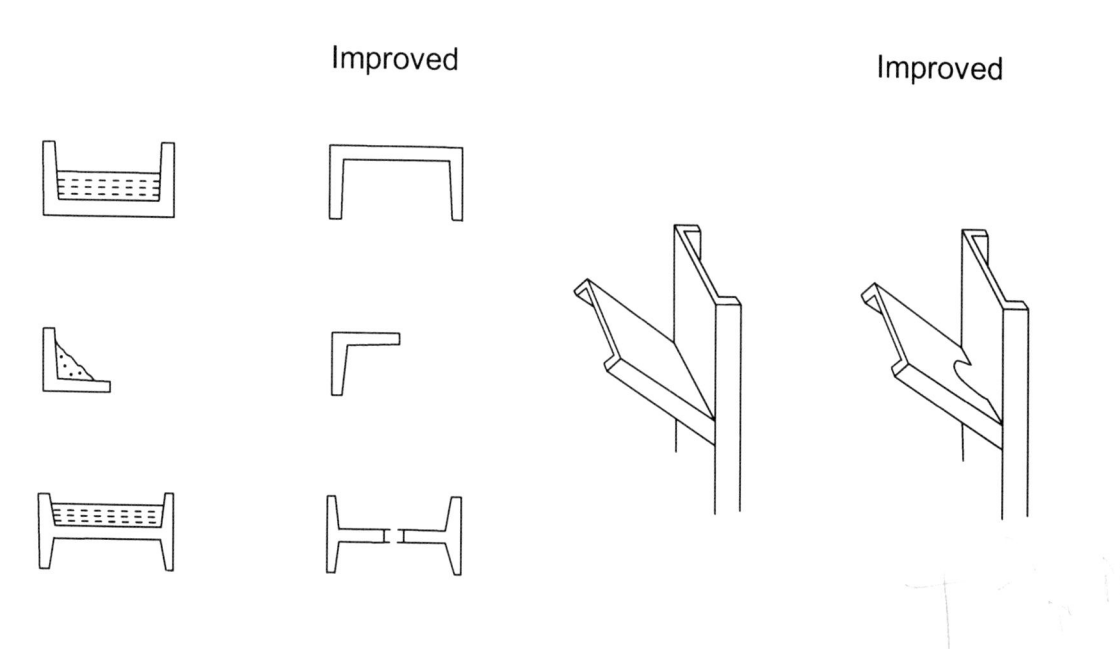

Figure 5.8: Improving corrosion resistance by detailing

- In addition, where there are bolted lap joints the bolts should be at or near the minimum pitch (i.e. 75 mm to 100 mm).

- Otherwise slight gaps in the plate contact occur between bolts and these gaps admit water thus allowing corrosion to take place.

Comment

A pinned connection is not uncommon in buildings (used for bracing mainly). Usually there is no fatigue problem, but as these last two cases have shown, it is difficult to inspect these connections for corrosion. Thus it is a good idea to make stainless steel if possible (with isolation washers fitted to prevent dissimilar corrosion of the steel).

5.3 Antioch School, California, USA, 1980.

[Mast (1980), Kaminetzky (1991)]

This failure occurred in the roof of the auditorium/gymnasium building of the school. It was constructed in 1959. The roof consisted of 350 mm deep precast prestressed lightweight concrete double tees (DT) spanning 12 m between simple supports. In order to reduce the total construction depth the ends of the double tees were "dapped" (i.e., undercut). As was the common construction practice at the time, the ends of the double tee were welded to the supports.

Collapse

The auditorium roof suddenly collapsed in 1980, (i.e., 21 years after construction). There was no unusual loading at the time. Some concrete spalling was the only warning of collapse. There were no injuries.

Investigation

It was realized that the connection detail at the ends of the double tees was the weak link in the construction. The connection is shown in Figure 5.9. The welding meant that as the double tee tried to contract as a result of concrete shrinkage, creep (it was prestressed so there was horizontal creep), and temperature drop, it was restrained. Thus tensile stresses developed over the depth of the section. A crack developed as a result of the lack of provision for movement.

Cracks are not usually a problem if they are controlled by reinforcement. The double tee lacked such reinforcement. Thus there was nothing to control the propagation of such a crack. Collapse eventually resulted due to this inadequate reinforcement at the dapped end of the double tee.

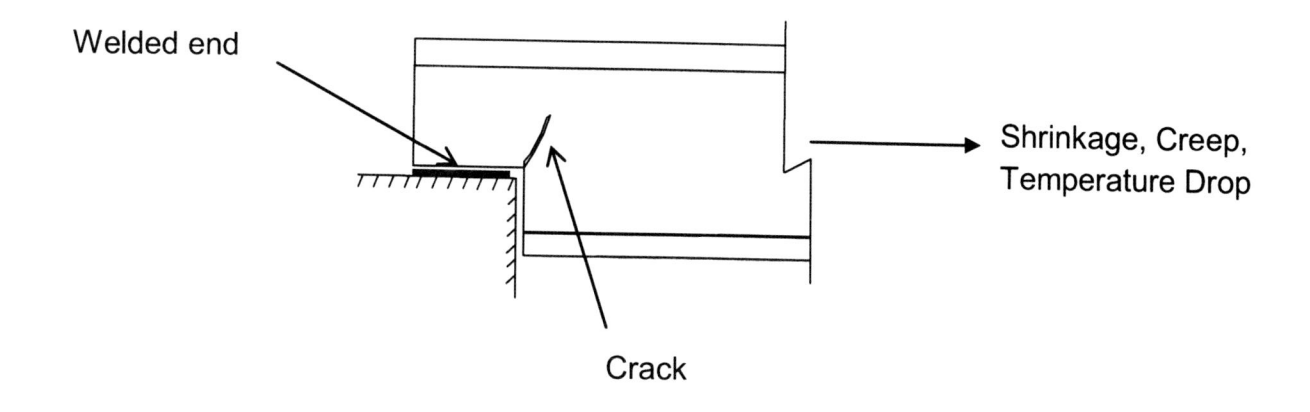

Figure 5.9: DT support connection

The preferred detail is shown in Figure 5.10.

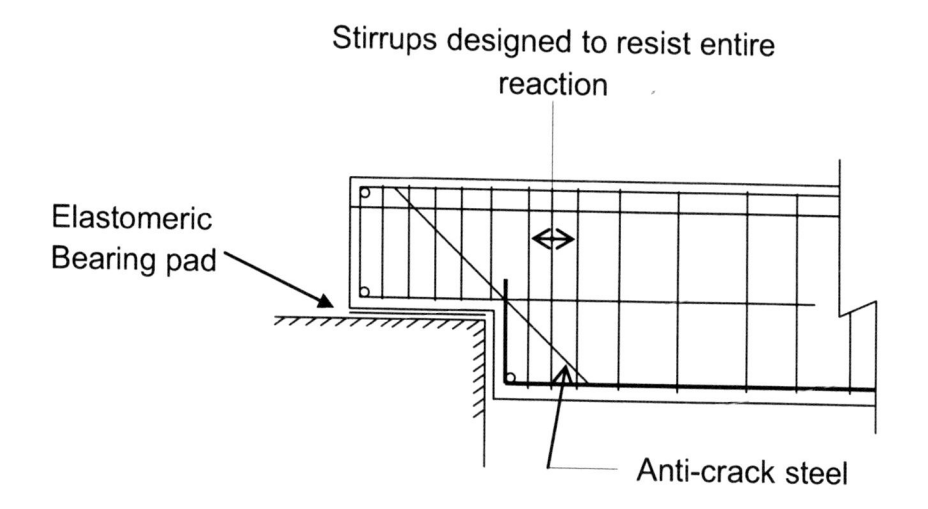

Figure 5.10: Preferred detail at DT support

Lessons

- Avoid welding <u>both</u> ends of prestressed members. Bearing-pads should be provided.*

- Avoid undercutting prestressed DT as it is expensive and difficult to accommodate the required reinforcement.

- Ensure that if the concrete cracks there is sufficient reinforcement present to avoid collapse.

*Note: The Author worked in the offices of Jan Bobrowski & Partners in the late 1980s. They designed a great many precast frames mainly for Grandstands etc in various parts of the world. It was the practice of that office to <u>always</u> weld both ends of a double tee, and to then completely build-it into the supports, with a minimum bearing of 100 mm, so the result was a monolithic beam. In addition, double tees were never allowed to be undercut. As well as improving the progressive collapse resistance of the structure, this encouragement of monolithic action improved the temporary stability too. There were no reported problems due to restraint. Thus the method recommended above is not the only satisfactory one. It appears there is no middle ground with this problem, either fully release or fully build in. And if you build-in, do not undercut.

5.4 Ynys-y-Gwas Bridge, West Glamorgan, Wales, UK, 1985.

[Woodward & Williams (1988), Discussion to Woodward & Williams (1989)]

This road bridge consisted of a single simply supported span of 18.3 m over a narrow river. It was built in 1953 using short precast I-beam segments post-tensioned together longitudinally and transversely. The longitudinal joint between each segment was 25 mm and was filled with mortar. The longitudinal post-tensioning tendons were placed in ducts cast into the precast I-beams. At the end supports the beams rested on bridge bearings made from building paper. A cross-section is shown in Figure 5.11 below.

Collapse

At about 7 am on 4th December 1985 the bridge suddenly, and without warning, collapsed under its self-weight by hinging at mid-span. The edge beams and parapets remained in place. There were no injuries or witnesses.

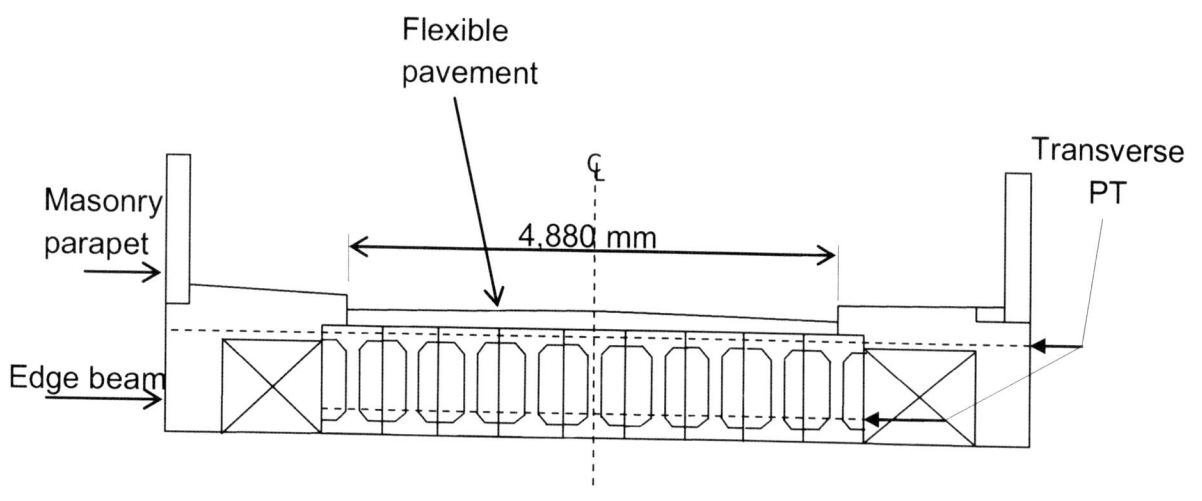

Figure 5.11: Cross-section of bridge

Investigation

The bridge had been inspected 10 times during its 32-year life. Its most recent inspection had been six months before the collapse. These inspections had been purely visual. However, there was no evidence reported of any distress (e.g., rust stains, cracking, spalling, or deflection) before failure.

Upon inspection of the debris, corrosion was evident in tendons at the I-beam segment joints only (both longitudinal and transverse). The tendons had lost 80-90% of their area over a few millimeters. The corrosion had been caused mostly by road salt (used to melt ice in winter).

Chlorides were found on the surface of the concrete (maximum 1% by mass of cement) and in the mortar (maximum 2%), but not in the post-tensioning grout. Most codes limit the total chloride-ion content in reinforced concrete to less than 0.4% by mass of cement. The UK's Building Research Establishment consider < 0.4% "low risk", 0.4-1% "medium risk", and > 1% "high risk".

Drillings showed a decreasing chloride concentration with depth. This meant these chlorides were probably not present during construction. It was reasoned that water containing these salts percolated downwards from the roadway. The mortar was found to be highly permeable (100 to 1000 times more permeable than the concrete).

There was little rust staining as the corrosion was highly localized so there was little corrosion product. There was no cracking or spalling of concrete as corrosion products were mostly magnetite (Fe_3O_4 known as black rust) which is not expansive. (Fe_2O_3 is "red rust").

Several things increased the vulnerability of the tendons:

- Absence of top RC slab;
- Use of cardboard sleeves at segment joints instead of metal ones;
- Use of improper bearings meant increased friction;
- Damp environment provided by river beneath.

It is likely that a heavy load passed over the bridge seconds before the collapse, and that acted as the trigger.

Lessons

- Corrosion of grouted tendons can reach an advanced stage without any external evidence.

- Avoid adding any chloride to the mix. Calcium chloride was a popular additive in the past but has been effectively banned (e.g., in the UK since 1985) as chlorides promote corrosion.

- Avoid segmental construction with mortar joints: epoxy joints between match-cast segments are preferred.

- A purely visual inspection of a structure is not enough if there is a risk of corrosion caused by chlorides. Ensure proper sampling by drilling.

- Avoid large statically determinate structures.

- Ensure there is redundancy: the structure should be statically indeterminate and strong enough to ensure that the introduction of a single degree of freedom does not result in collapse.

- Experience has shown that the most effective technique for detecting corrosion in bonded tendons is invasive coring or drilling to determine carbonation levels, chlorides etc.

5.5 Schoharie Creek River Bridge, New York, USA, 1987.

[Levy & Salvadori (1992), Schlager (1994), Feld & Carper (1997), Various (1998), Wearne (2000)]

The bridge, known as the "Thruway", was constructed in 1954. It consisted of five simply supported spans of steel girders supporting concrete deck. Each support consisted of two concrete piers resting on a concrete plinth. The plinths had shallow footings (the rock was about 15 m below river). The arrangement is shown in Figure 5.12.

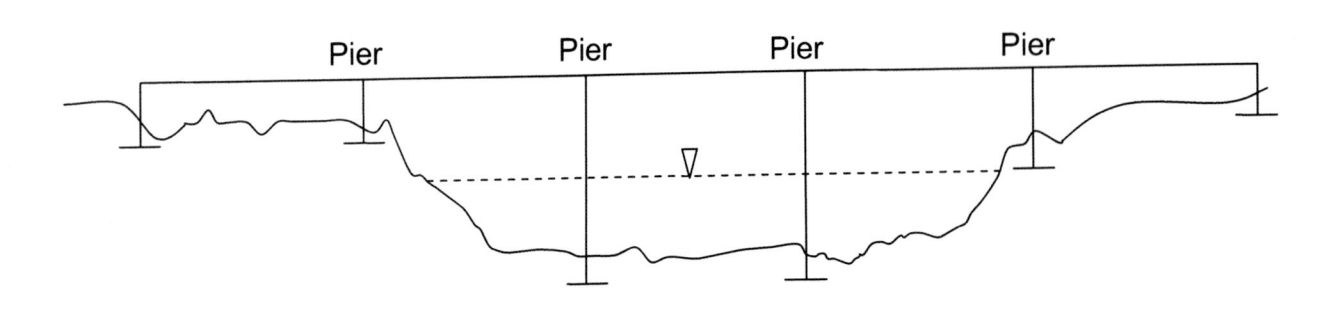

Figure 5.12: Schoharie Creek Bridge

Collapse

On April 5th, there was heavy rain in the area (about 280 mm in 24 hours) and this added to the melted snow from nearby mountains ensuring that the river was not a "creek" but more like a raging torrent.

Suddenly, the third set of piers from the west collapsed (pier 3). Two spans collapsed immediately. Within 90 minutes the second pier from west (pier 2) and another span of the bridge also collapsed. The power of the rushing waters was so great that one vehicle was carried 1,500 m downstream. Ten people were killed.

Investigation

New York State required all bridges to be inspected **annually,** with inspection of underwater elements every five years. Although the yearly inspections were done, the underwater portion of the bridge was never inspected. Until 1985 Schoharie Creek was not considered "deep water" so it was felt an underwater inspection was unnecessary.

The first underwater inspection was due a few weeks after the bridge collapsed. The report of the inspection in indicated that "rip-rap" (large angular stones weighing about 125 kg each) placed earlier to protect the foundations, was no longer visible from above and so should be replaced. It was scheduled to be replaced in 1981.

However, in 1980, a non-engineer state employee viewed the plinths from the shore when the river was low. As there were no problems evident he decided the rip-rap was not necessary. Nobody else was consulted and no questions were asked. No new rip-rap was placed in 1981. In fact, it was unclear as to how much rip-rap had been placed originally to protect the foundations. Design drawing were unclear, some indicating that the cofferdam used to construct the piers was to be backfilled with rip-rap to full depth, others indicating rip-rap at the top only. There was no inspection to confirm the amount placed.

It was found after the collapse that the hole scoured out under pier 3 was up to 3m deep. The plinths supporting the piers simply tipped upstream into the holes scoured out by the fast moving water.

It was noted that had the girders been continuous instead of simply supported, the failure of one pier might not have caused total collapse or at least given warning.

Lessons

- Avoid large statically determinate structures.
- Avoid untrained and unqualified personnel making important civil engineering decisions.
- Scour is the single most common reason for failure of bridge structures. To avoid it:
    - Use piles for foundations in river, or
    - Protect shallow bridge foundations in river with rip-rap.

Additional Information

The remaining bridges of similar design were retrofitted by adding rip-rap (where necessary) and post-tensioning the plinth so that even if a hole opened up under one side of the plinth, it would not fall into it.

Before the collapse the police had closed several bridges crossing Schoharie Creek. Unfortunately, the Thruway Bridge over Schoharie Creek was outside their jurisdiction.

In July 1990 a bridge over the Inn River in Austria, suffered the loss of a pier under similar circumstances. However the bridge was continuous and warning was given by large deflections. There was no loss of life.

In August 2009 a pier of a rail bridge over an estuary in Ireland was scoured away resulting in the collapse of two spans of this statically determinate structure. It happened as a train was crossing the bridge. Luckily nobody was injured. The accident report describes the gap in inspector's knowledge due to staff movements, resulting in the failure to inspect adequately.

5.6 Piper's Row car park, Wolverhampton, UK, 1997

[Wood (2003a), Wood (2003b)]

The car park was a 4-storey reinforced concrete building with reinforced concrete flat plates 225 mm thick. It was constructed in 1965 by the then popular lift-slab technique. In the lift-slab construction technique, all slabs are cast one on top of another at ground level (separated by a membrane) then lifted up precast concrete or steel columns (in this case the columns were precast concrete), using hydraulic jacks. At their final level they are welded in place using steel wedges. The joint was then grouted. About 80 car parks in the UK were constructed using this technique.

Collapse

At about 3.15 am on 21st March 1997 part of the roof slab collapsed without warning. The collapse progressed horizontally: a 15 m x 15 m section fell. No progression of the collapse took place vertically. After the collapse, the level 4 slab rested on that of level 3. This may have been due to the presence of vertical mullions around the perimeter (nominally not part of the vertical load bearing structure). Alternatively the level 3 slab may have had sufficient ultimate resistance to allow the support of the level 4 slab.

Failure occurred around 8 columns. Columns remained in place in most cases, with a small portion of slab on top. See Figure 5.13. Some patch repairs were carried out in 1996 and a full investigation was recommended then. It had not started by the time of the collapse. Thus much of the upper and the 3rd floor were already cordoned-off and there were no injuries.

Investigation

Clearly the connections between the columns and the slab had failed by punching. The top surface of the concrete was only lightly carbonated (1-2 mm): there was no obvious corrosion.

However, many factors combined to reduce the punching strength:

1. The Codes in use at the time in the UK (CP114: 1957) did not address punching properly (not enough punching reinforcement was specified).

2. The concrete quality was highly variable.

3. Top reinforcement in the slab was placed low.

4. The wedges were placed at uneven heights relative to adjacent columns causing some column connections to be loaded 40% more than assumed.

5. The deflection and creep of the slab created areas of ponding on the surface. The water seeped into the slab at these ponds. Once it froze, it damaged the concrete.

Friable, degraded material was found up to a depth of 100 mm. This degradation was sufficient to destroy most of the bond between the steel reinforcement and the concrete further reducing the punching strength. (There is little reduction in punching strength until the bond of the top steel is destroyed.)

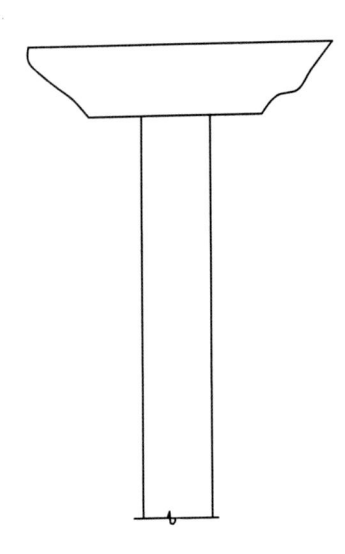

Figure 5.13: Undamaged column at Piper's Row

Lessons

- The factor of safety against punching can easily be reduced as a result of poor concrete, etc.
- Punching is a brittle mode of failure (as are conventional shear failures) so the factor of safety against it should be even higher.
- The passage of time can mean deterioration takes place in the slab. The only visible sign may be water leakage.
- The top deck slab of a car park should be protected with a proper waterproof membrane.
- If a flat slab which does not have bottom bars deteriorates, progressive collapse can follow.

- Providing only ties may just ensure that portions of the structure adjacent to that which collapses are also dragged down.

"...detailing bottom steel through columns ensures a ductile punching failure will not spread. Without ductility a much higher factor of safety is needed."

Prof. J.G.M. Wood (investigator for Piper's Row)

5.7 De La Concorde Overpass, Montreal, Canada, 2007.

[Commission (2007), Wood (2008), Delatte (2008)]

This road bridge carried three lanes in each direction. It was constructed in 1970. It was a concrete bridge consisting of a precast main span (27.5 m) supported by *in situ* cantilevers at each end using a half-joint. There was an expansion joint above each half-joint. The bridge was visually inspected every two years. (The inspection reports usually said there was nothing wrong except that the expansion joints had failed. This is a common occurrence on bridges). The bridge is shown in Figure 5.14 and the half-joint in Figure 5.15.

Collapse

At 9:30 am on 30 September 2006 a passer-by reported concrete spalling from the bridge. At 11:45 am one of the bridge maintenance personnel (not an engineer) responded. He visually inspected bridge but found nothing abnormal. He recommended that the bridge remain open but that later inspection by a qualified person should take place. At 12:30 pm the bridge collapsed suddenly. Failure began at the right-hand half-joint of bridge. Five people were killed.

Maintenance

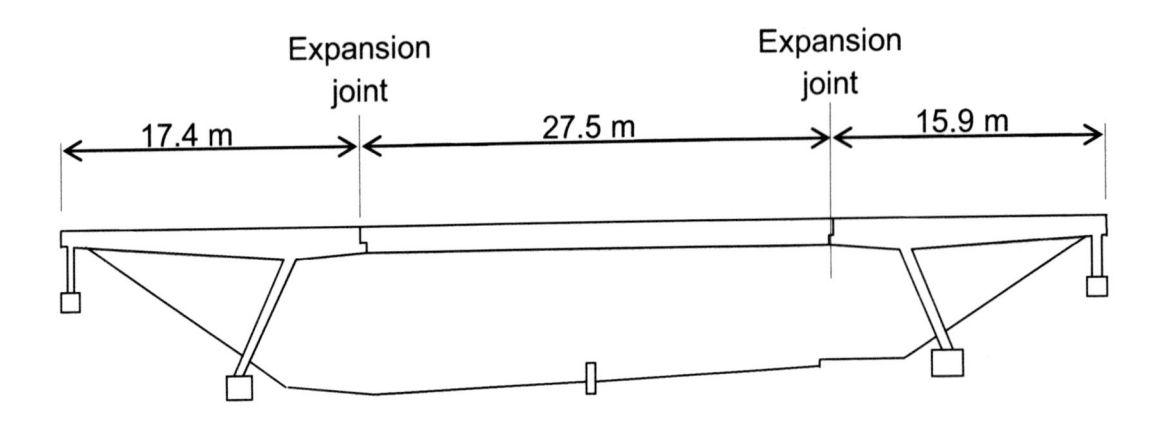

Figure 4.14: De La Concorde Overpass Bridge

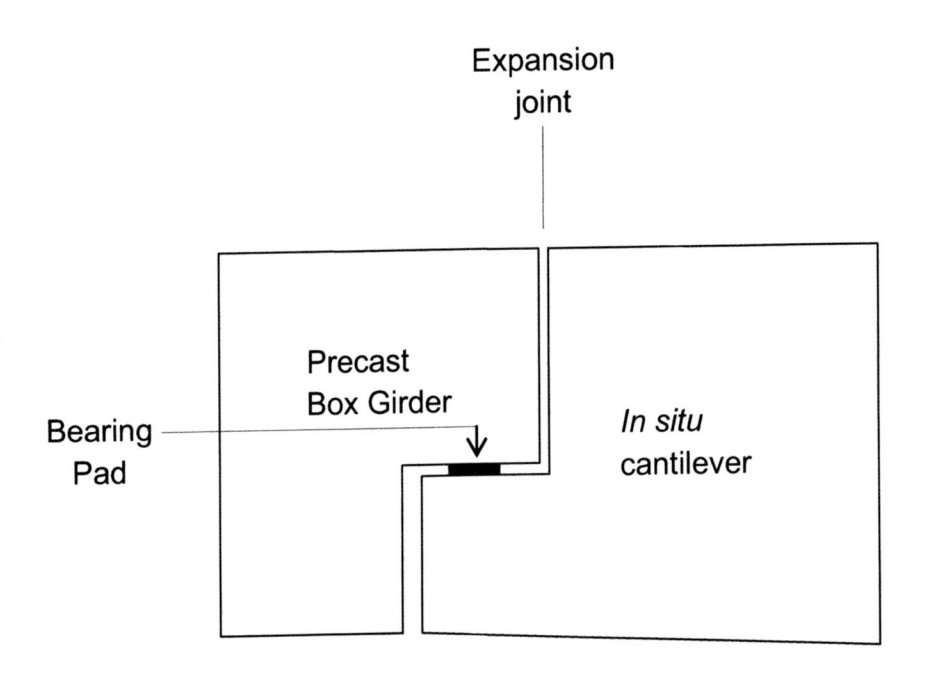

Figure 5.15: Detail at expansion joint

Investigation

The original construction contractor was inexperienced, but as the price quoted was the lowest, was awarded the job. Construction workmanship was poor; concrete quality inconsistent. Thus there was little resistance to attack from de-icing salts and frost especially at top of bridge (where the least compacted, and so poorest, concrete can be expected to be). There was no waterproofing specified to top of the deck (a liquid membrane was installed 1992).

Design against shear failure complied with rules as in 1970, but was unacceptable to current standards. The figure below, Figure 5.16, shows the design intent for the detail of the reinforcement in the cantilever next to half- joint. (It is now known that vertical, diagonal and main bar reinforcement should be properly anchored).

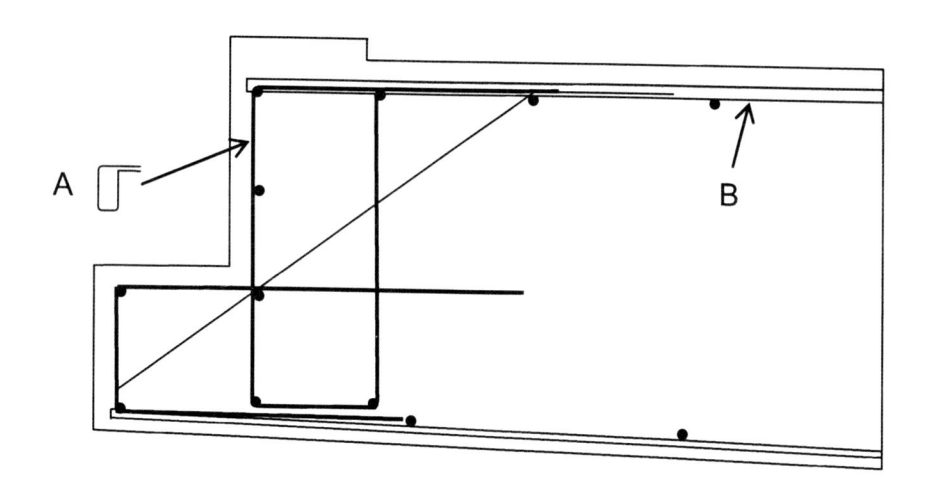

Figure 5.16: As-designed

The intent was to lap the top bar 'B' in the horizontal plane with the horizontal part of the lower bar 'A'. However, the contractor did not carry out the design intent. The figure below shows the detail actually provided (Figure 5.17).

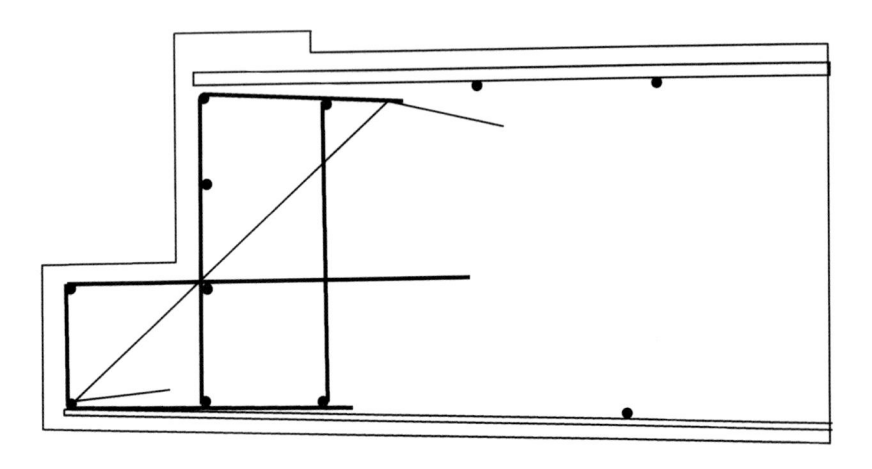

Figure 5.17: As-built

A lap was provided but in the **vertical** plane. Any lap requires struts within the concrete to be developed. These struts require to be equilibrated. If no reinforcement is present to restrain the lap from its tendency to separate then it is the tensile strength of the concrete that must equilibrate the struts. See Figure 5.18.

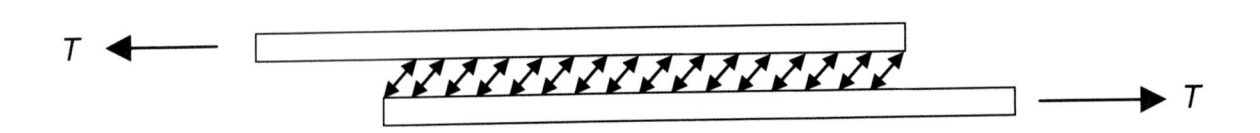

Figure 5.18: Reinforcement lap

So a lap requires tensile strength from the concrete in the vertical direction, just at the place it is weakest, at the top of the section.

In the design of the cantilever assumed it was a "slab" rather than a "beam" (this was allowed by codes at the time). Thus, according to the code rules, no shear reinforcement was provided (not even nominal). As well as adding to the shear capacity, the reinforcement would, if conventionally detailed, have enclosed the lap. Thus it would have made the failure more ductile even if there was insufficient link reinforcement to completely prevent the failure. The load at which shear failure is likely to occur is particularly difficult to estimate if no links are provided as it depends greatly on the tensile strength of the concrete.

Repairs to the expansion joint and the concrete at the top of cantilevers were carried out in 1993. The opportunity to discover reinforcement detailing errors were missed. In 2004 inspectors expressed concern at cracking and efflorescence on the face of the cantilever where failure initiated. However inspecting engineers considered that RC was ductile and so any failure would happen slowly. No reference to the original reinforcement detailing drawings was made. This would have alerted inspectors that there was a serious problem.

Load tests were done on laboratory specimens constructed using the same details as the as-built connection. The failure was reproduced, the critical crack forming due to the poor lap detail and propagating due to the absence of links to eventually cause shear failure. See Figure 5.19.

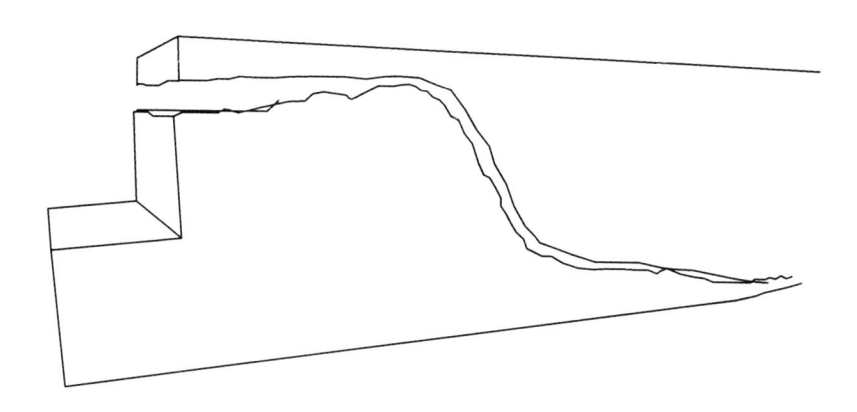

Figure 5.19: Laboratory tests reproduced failure mode

Lessons

- Brittle (shear) failure gives little warning (especially if no links have been provided).

- Insufficient or poorly detailed/placed reinforcement can make any RC structure insufficiently strong or brittle. (Expect all concrete failures to be brittle, except when the under reinforced flexural capacity is reached.)

- The definition of the flexural element as a "beam" or a "slab" can mean different detailing used.

- The half-joint detail is difficult to inspect and avoided in new bridges for that reason; (it also makes removal of bearings difficult).

- This detail vulnerable as expansion joint failure can result in salt containing water collecting on seat. (Much less salt was used in 1960s so designers may not have considered this a big risk.)

- The statically determinate main span meant collapse was inevitable once support at one end was lost.

- Inspection of RC structures should include a proper re-appraisal at least once in the life of the structure as materials deteriorate and codes change.

- Only then can older structures be properly assessed in the light of improving knowledge.

"Inspections should not be confined to searching for defects which may exist, but should include anticipating incipient problems. Thus inspections are performed in order to develop both preventative as well as corrective maintenance programs."

AASHTO *Manual for Condition Evaluation of Bridges*

The following is offered as a means of deciding which structures should have priority for maintenance. Of course, all should be maintained at some time.

Suggested Priority for Maintenance of Structures

1. Collapse would be especially catastrophic;

2. Statically determinate structures;

3. Designed to outdated codes of practice.

    E.g., before 1972 many concrete structures were designed to CP114; it was later realized that the shear and cover provisions were inadequate; before 2006 flat slabs designed to BS8110 required no bottom bars.

Comment:

It is well known that structures designed "to the bone" are likely to be more expensive to maintain. Construction and maintenance are usually considered separate by client as they are often paid for in different ways. Thus the client is often uninterested in any structural schemes that are more expensive to construct even if the "whole life" costs are lower. In addition, it has been observed that people are inclined to discount future events (Bazerman & Watkins 2008).

Chapter 6: Welding

"Steel is a wonderfully ductile material--as long as you don't try to connect it"

(Nigel Priestley, quoted in Feld & Carper 1997)

6.0   Introduction

[Dowling et al. (1988), Gordon (1988), Ashby & Jones (1998), McEvily (2002)]

Welding is an extremely efficient and elegant way of joining steel components. It improves corrosion resistance as the lines are cleaner and so don't trap dirt. The commonest welding method is arc welding. The melting point of steel is about 1,450°C the temperature of the arc is about 6,000°C.

Hot steel can dissolve lots of carbon. Cool steel cannot dissolve as much. As steel cools, the carbon precipitates out as a new carbon-rich material (called "iron carbide"). A fast cooling rate results is severe distortion of the iron crystal as the carbon attempts to precipitate out. The precipitate is then called "martensite". Welding involves rapid heating and cooling of the steel (the cooling rate can be as high as 1,000°C/sec). The risk is the formation of martensite (a hard, brittle form of steel). If martensite is present then the weld is likely to be weak and brittle.

The parent metal of Grade 43 and 50 steels must have a Charpy of about 27 J/m^2 at minus 20^{0}C. The welded connection may be more brittle. Avoid welding materials with high carbon contents. Other alloying elements can make things worse too. We can define a "Carbon 'Equivalent' (CE) = C + Mn/6 + (Cr +Mo +V)/5 + (Ni + Cu)/15

"Weldability" is determined by original steels' CE and the cooling rate. EC3 cautions against welding structural steel if the Carbon Equivalent is more than about 0.4%. Thus there is no problem with normal grades of structural steel (e.g., Grade 43, 50).

For example, one welding technique is Manual Metal Arc (MMA) welding (see Figure 6.1):

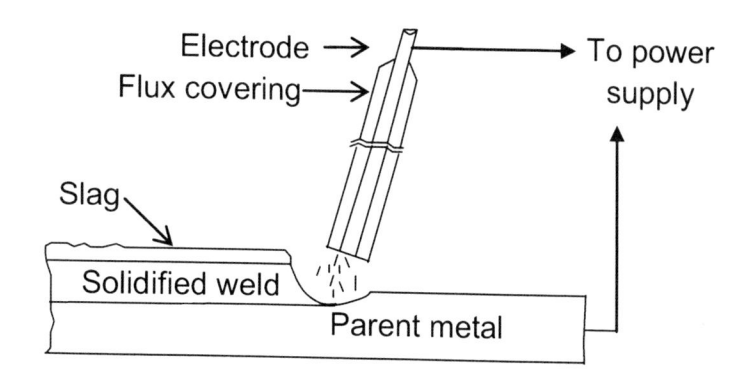

Figure 6.1: Manual Metal Arc welding (MMA)

The flux coating allows an inert gas to form which protects the weld from atmospheric contaminants like hydrogen, which causes embrittlement of steel.

Generally thick pieces of steel should be avoided as they make the cooling rate faster, and so the risk of martensite formation is higher. A small weld on a thick plate is an especially poor detail (as the cooling rate is high). Similarly attempting to weld a thin piece to a thick piece is troublesome.

When a weld cools it shrinks causing distortion of the welded steel. An experienced welder can minimize this distortion. Thus the term "workmanship" here refers not just to the quality of the weld itself. If this distortion is restrained tensile stresses build up; (the stresses in the heating phase are compressive).

Triaxial tensile stresses reduce the ductility of the steel (as there is less shear to cause the dislocations to move). Again the problem is worse for thicker plates.

Before welding, ensure that the surfaces to be welded fit together properly and have adequate chamfering so that proper penetration is possible.

Usually site welding is avoided if possible. It is both more expensive (e.g., special staging has to be erected) and likely to be of lower quality. Shop welding is preferred and site connections are usually bolted.

Lamellar tearing

Steel is not perfectly isotropic, i.e., its strength varies depending on the direction of the load. Rolling causes "slag" (consisting of limestone/silica compounds), which is present to various degrees in all rolled steel, to align with the rolling direction and so form a zone of weakness at that point. This plane of weakness may be revealed when tension is applied perpendicular to this point, e.g., by the restrained shrinkage of a weld. See Figure 6.2.

The thicker the plate, the worse is the effect. Ultrasonic testing can detect this type of defect. Failure due to lamellar tearing usually occurs during fabrication/construction. The following cases are from Kaminetzky (1991): (1) Performing arts Centre, El Paso, (2) the 100-storey Hancock Centre, Chicago, and (3) the twin 52-storey towers in Los Angeles. Welding defects are usually only repaired at great expense (e.g., $400,000 in case (3) above).

About 1% of the volume of steel is occupied by inclusions. These are slag or sulphur-containing compounds. These inclusions are rarely uniformly distributed. When steel is cast into ingots the inclusions collect around the centre-line and near the top of the casting. Using low sulphur raw materials (i.e., sulphur not exceeding 0.01%) and adding calcium, reduces the inclusions to give "clean" steels. "Vacuum de-gassed" steel is available (containing very low impurities and very low sulphur) to reduce the problem, but costs more. It is used mainly in the offshore industry.

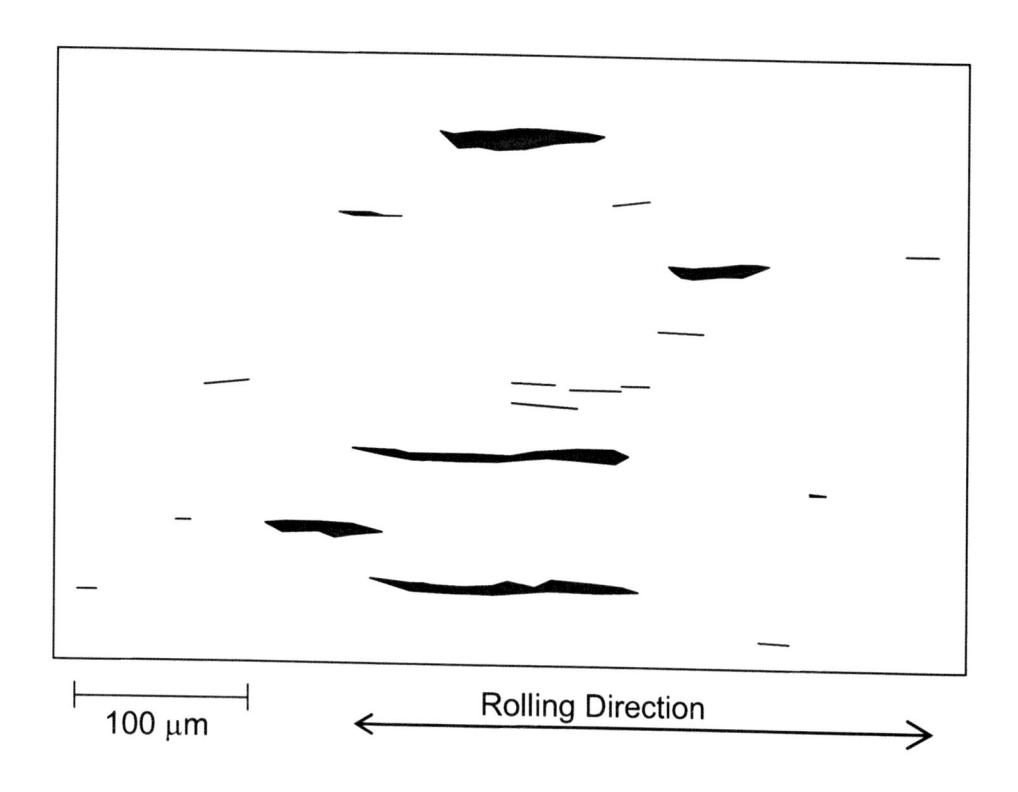

Figure 6.2: Laminations in rolled steel

The following illustration (from James 2000) compares the properties of a typical rolled steel 50 mm thick plate in both directions.

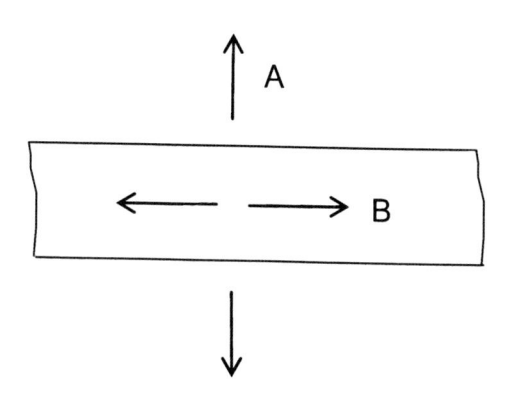

Table 6.1 below indicates that, comparing the two directions, there is a loss of strength of about 33%, and of ductility of about 75%.

Table 6.1: effect of direction of rolling

Direction	Ultimate Tensile Stress (N/mm^2)	Yield Stress (N/mm^2)	% Elongation
A	319.2	None	6
B	476	298.9	25

It is suggested (Kaminetzky 1991) that the following type of improvement is made to detailing to decrease the susceptibility to lamellar tearing. See Figure 6.3:

Susceptible detail · Improved detail

⟷ Direction of rolling
T   Tension due to restraint

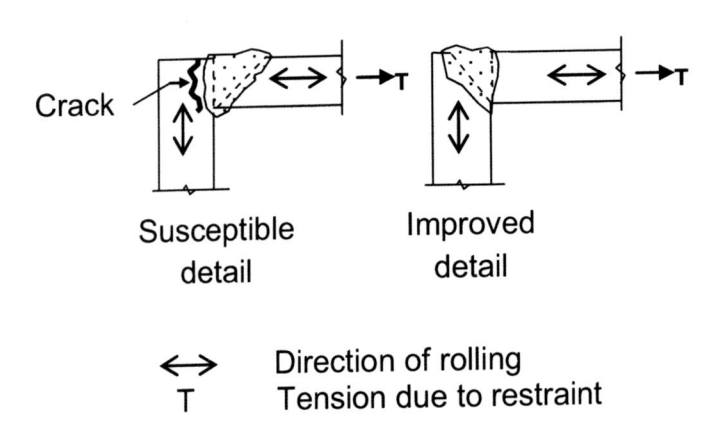

Figure 6.3: Detailing to improve resistance to lamellar tearing

The following checks are recommended for all welded structures (Ratay 2010):
- Prior to welding steel plates of 38 mm or greater the plate should be ultrasonically tested for imperfections and laminations;

- All complete joint penetration welds should be 100% tested using ultrasonic or radiographic methods;

- Fillet welds should be inspected using magnetic particle or dye-penetration methods. 10-15% of lineal dimension of the weld should be checked.

- Case Studies in Chapter 6:
    - King's Street Bridge, Melbourne, 1962.
    - Brooklyn College, New York, 1971.
    - Alexander Kielland, Norway, 1980.
    - Failure of SMRF during Northridge earthquake, 1994.

6.1   King's Street Bridge, Melbourne, Australia, 1962.

[Dowling et al. (1988), Francis (1989), Schlager (1994), Shepherd & Frost (1995), Feld & Carper (1997), Ashby & Jones (1998), McEvily (2002)]

This bridge consisted of 30 simply supported spans, supporting a highway and a railway over the Yarra River. The spans varied up to about 49 m. Each span consisted of four welded steel plate girders supporting an RC deck. It was completed in April 1961. The specification limited flange plates to a maximum of 25 mm thick in an effort to prevent brittle fracture.

Collapse

One cold winter morning in July 1962 (temperatures were about -1°C), the end span (30 m) failed suddenly under the weight of a vehicle. The four parallel girders had failed. It sagged only about 300 mm as fortunately there were retaining walls underneath and they prevented total collapse of the bridge. There were no injuries.

Investigation

The vehicle that triggered the collapse weighed 45 tonnes, well under the load limit for this bridge. Most of the other girders which did not collapse were found to have cracks. The girders were of medium high-strength steel with yield strength about 30% more than mild steel. The CE was about 0.6%. The welding found to be difficult; however no expert was consulted.

(a)

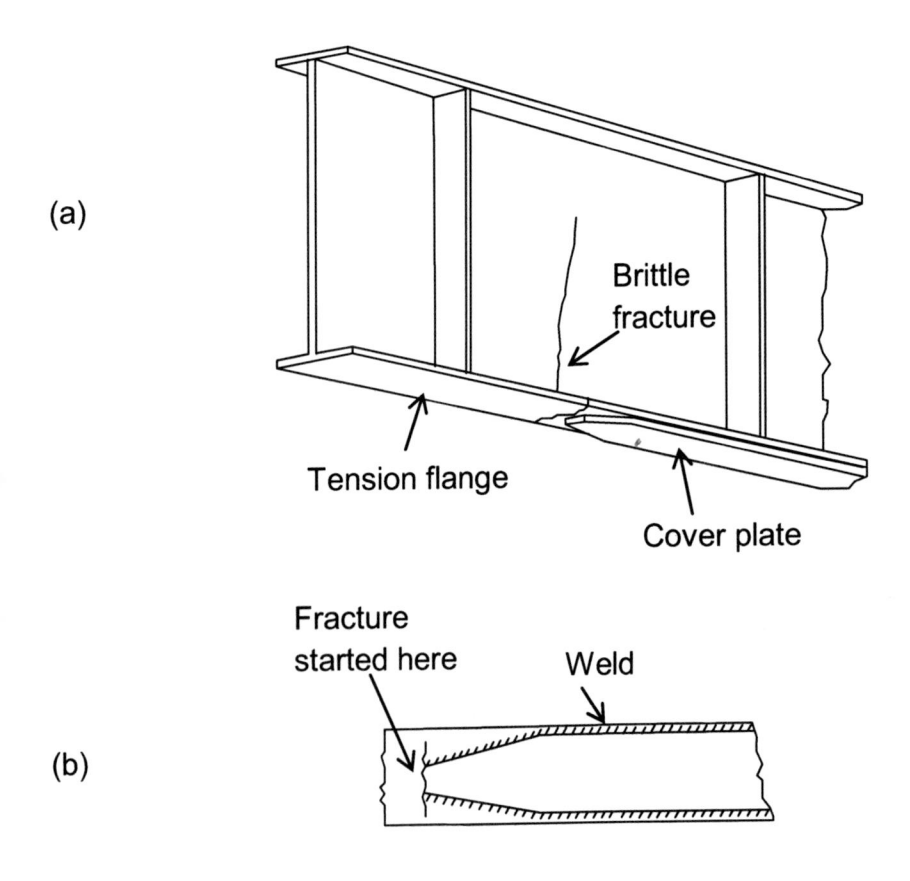

(b)

Figure 6.4: Details of welding at King's Street Bridge (a,b)

(c)

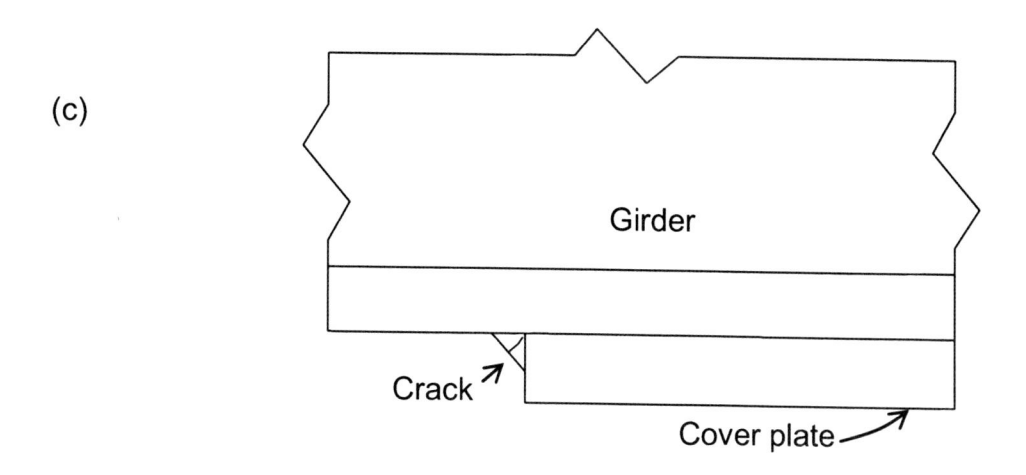

Figure 6.4: Details of welding at King's Street Bridge (c)

Cover plates were welded to the bottom flanges over the central part of the span (using a 5 mm fillet weld). See Figure 6.4 (a) and (b). Manual welding was used. Due to incorrect storage, the welding electrodes were damp. The cracks which led to the failure all started at the ends of the cover plates. The ends of the cover plates were welded **last** after the rest of the welding had been completed. The weld at end of cover plate tried to shrink as it cooled. (Imagine the weld shown in Figure 6.4 (c) above trying to pull cover plate towards it). It was not able to shrink freely, so high tensile stresses were created. Gravity loads ensured the zone around the end weld was always in tension. Thus cracking developed.

The welders should not have risked the occurrence of a transverse crack such as this. A longitudinal crack poses much less of a threat. Inspectors failed to notice these transverse cracks but diligently asked for several longitudinally cracked welds to be redone. That the cracks began in the factory was clear from the presence of priming paint in some cracks. The presence of traces of subsequent layers of paint applied after erection showed that the cracks extended quickly. In one case the crack was found to have propagated until the entire bottom flange and half the depth of the web. The shock caused by the passage of the vehicle on a cold morning was sufficient to start the final process of fracture.

Lessons

- The Carbon Equivalent was high and no special precautions were taken (e.g., pre-heating). If the CE is high (0.3% to 0.45%) then pre-heating of the weld to about 150-200°C is often considered necessary. For a CE above 0.45% pre-heating to up to 370°C is desirable. As well as slowing down rate of cooling, hydrogen is allowed to escape.

- Damp electrodes allowed hydrogen to seep from the atmosphere into the molten metal and embrittle it.

- Welds should be allowed to shrink when cooling. If not tensions are introduced. This is especially true of fillet welds.

- Low ambient temperatures make brittle failure more likely.

- If possible avoid cover plates. Use box sections rather than thicker plates.

- If the detail cannot be avoided then it is better to use wide, thin cover plates (wider than the flange) and so smaller welds. Intermittent welds are preferable in this case. Avoid over-welding (having welds unnecessarily large). Start at one end and move to other end.

The preferred detail is shown in Figure 6.5:

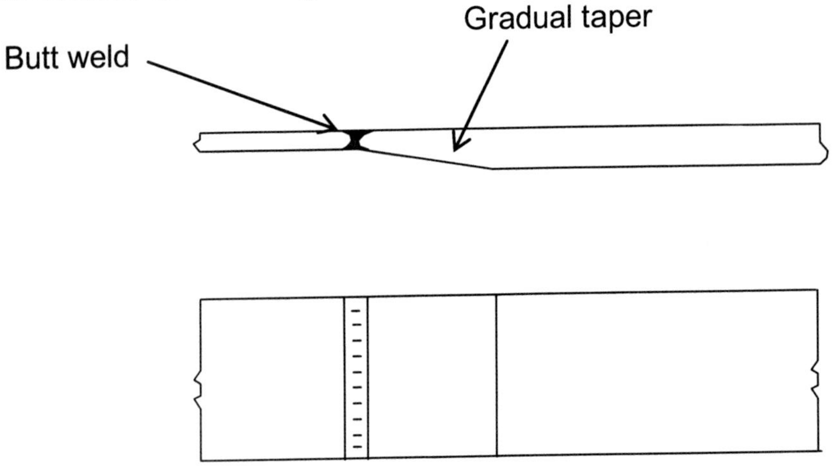

Figure 6.5: Preferred detail

Aftermath

The bridge was repaired by post-tensioning. Concrete blocks were cast at the ends of each span and high tensile steel rods were tensioned against them. The level of compression introduced prevented the steel from going into tension even under worst loading.

6.2   Brooklyn College, New York, USA, 1971.

[Kaminetzky (1991)]

This case concerns a new six-storey college building under construction. Welded plate girders were fabricated from A441 steel (similar to Grade 50 steel, f_y = 345 N/mm^2) to transfer loads from five storeys above. The structural arrangement is shown in Figure 6.6. An elevation of a typical plate girder is shown in Figure 6.7.

Collapse

One cold December day, several girders failed when the structure was nearly complete. Cracks appeared in flanges of girder, adjacent to welds of diagonal members (the girders were unusual in that a portion of the web consisted of a "truss" to allow services to penetrate).

A crack began due to weld shrinkage (see Figure 6.8). The presence of inclusions resulted in the propagation of the crack (lamellar tearing). This resulted in brittle fracture as the crack propagated through the rest of the flange. (The flange was in tension as it was in a hogging moment region).

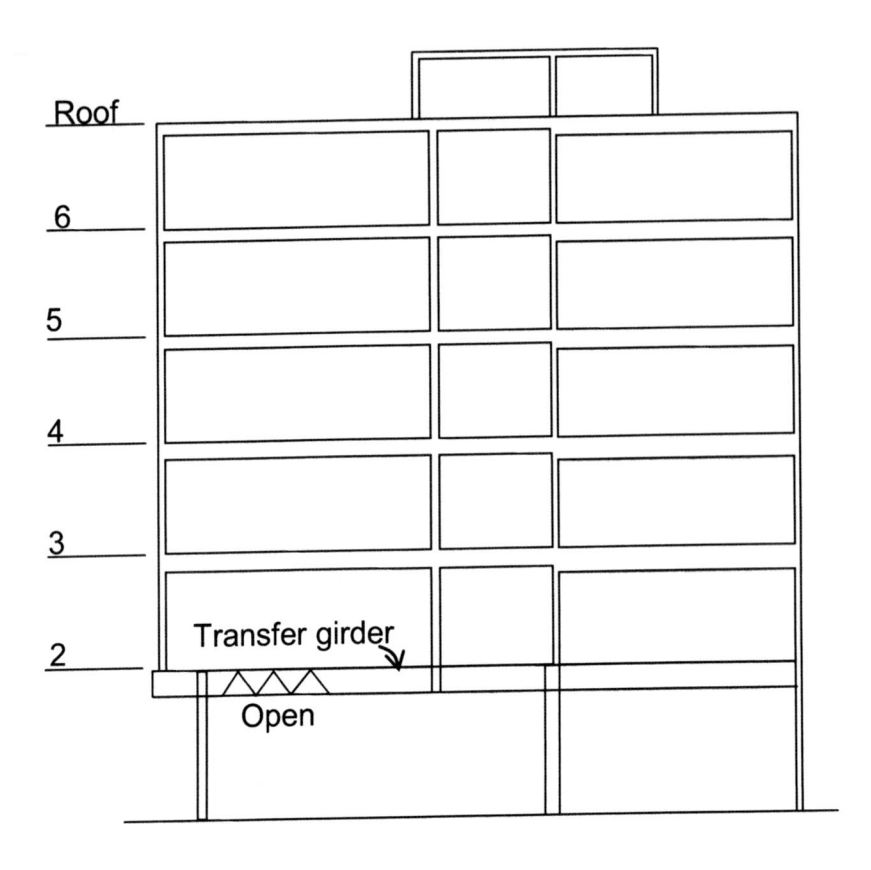

Figure 6.6: Cross-section through Brooklyn College Building

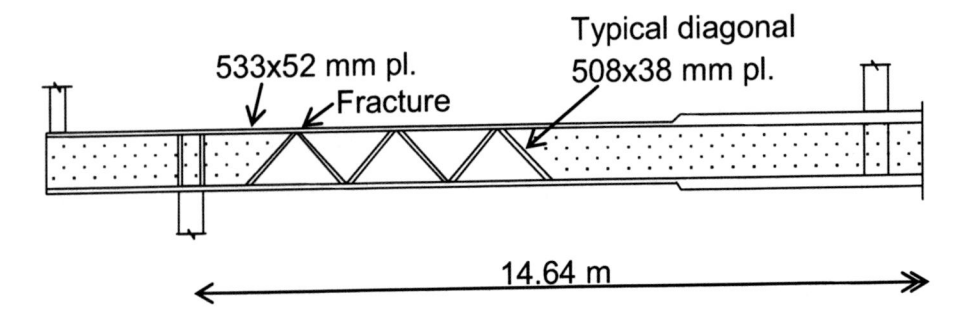

Figure 6.7: Elevation of typical plate girder

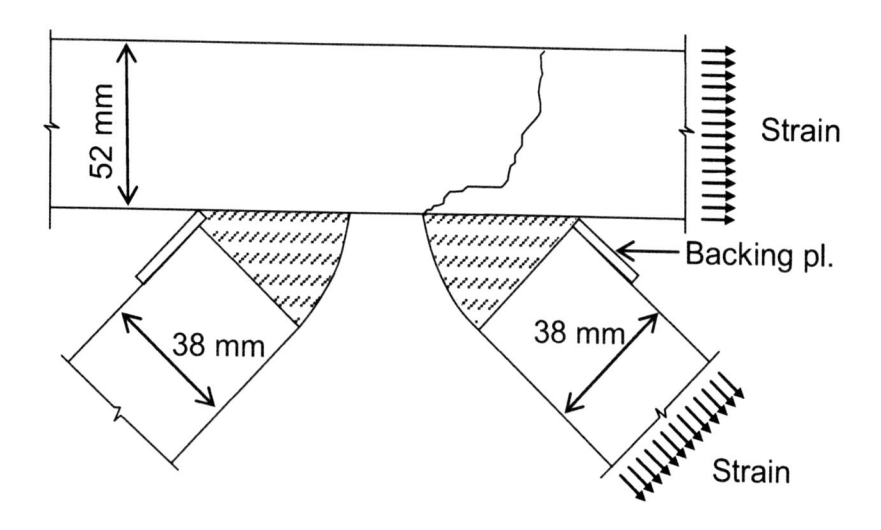

Figure 6.8: Connection between top flange and diagonals

Lessons

- Avoid using welds on thick pieces of steel.

- The risk of laminations being present increases with the thickness.

- Usually plates over about 38 mm thick are most affected, but plates as thin as 12 mm have been found with laminations.

- Specify ultra-sonic testing before fabrication, as well as post-welding inspection for cracks if must use details vulnerable to laminations.

- Research has shown that any moisture introduced into the weld worsens lamination failures.

- Preheat if necessary (say if using thick plates).

6.3 Alexander Kielland, Norway, 1980.

[Ross (1984), Gordon (1988), Petroski (1992), Schlager (1994), Ashby & Jones (1998), Wearne (2000), McEvily (2002), Jennings (2004)]

The Alexander Kielland was a five legged "semi-submersible" oil-rig. Controlled amounts of water flooded pontoons attached to ends of each leg so the rig could float with pontoons 20 m below water surface. The advantage of five legs was reduction in vertical motions caused by waves (recall that a minimum of three legs is needed for stability if all three can take tension).

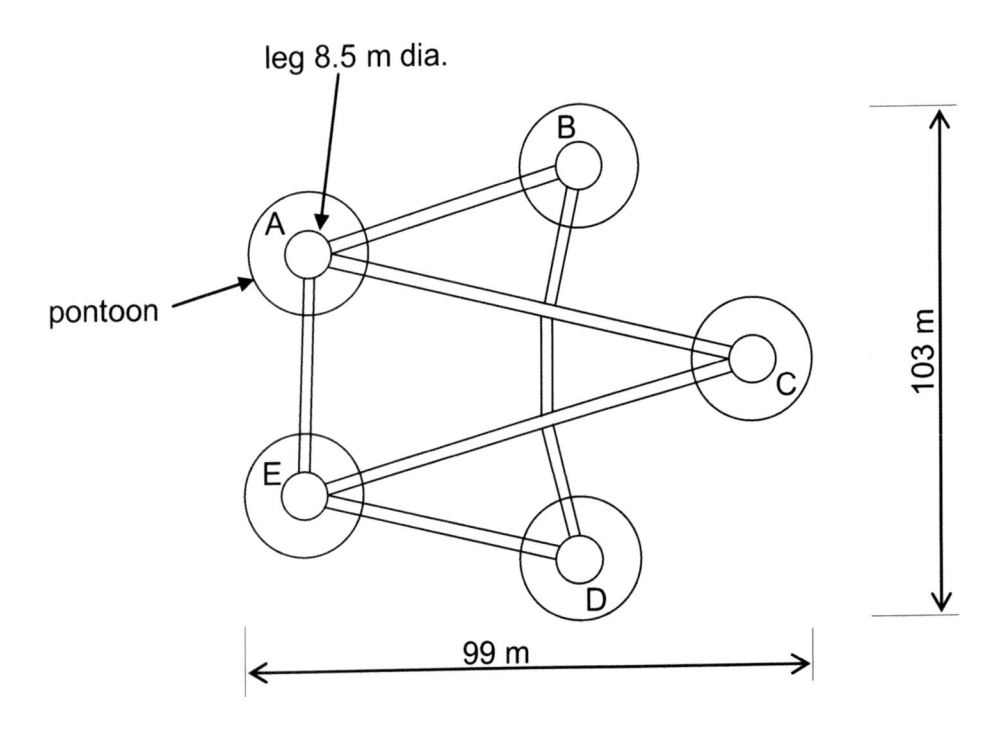

Figure 6.9: Plan of the Alexander Kielland support system

It was built in 1976 and was all-welded construction. A plan of the base of the rig is shown in Figure 6.9. Although built as an oil-rig, it was instead used as floating hotel in the North Sea. There was accommodation for up to 350 people.

Collapse

On the 27th of March 1980 there were 212 people aboard. The wind was blowing up to 20 m/s and waves were up to 10 m high. This storm should have been well within the capacity of the structure. Suddenly a loud crack was heard. The survivors saw that a leg of the rig had fallen off. The rig capsized shortly afterwards; 123 died.

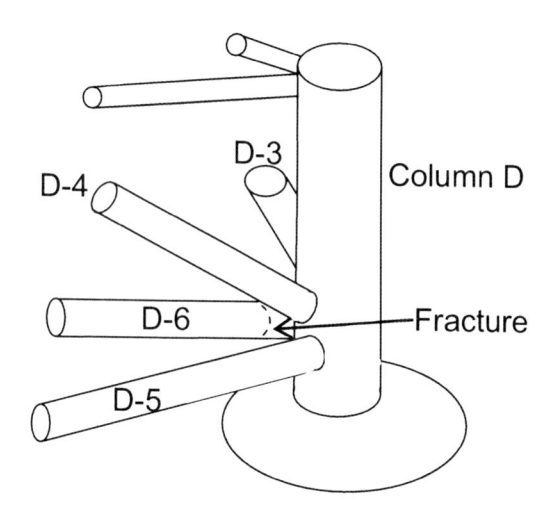

Figure 6.10: Column "D"

Investigation

The structure was designed for a one-hundred year storm. The leg that fell off was known as column "D" (see Figure 6.10). This separated leg was towed to Norway. Inspection of the end of a part of the brace "D-6" still attached to column showed fatigue cracking. This cracking seemed to start at a fillet weld. From this there was no doubt metal fatigue was to blame.

The structure was all-welded. Thus it was completely continuous. Cracking began at the fillet weld attaching a 325 mm dia. steel cylinder containing a hydrophone (used to aid positioning) to the underside of the 2.6 m diameter brace "D-6". As this weld was done after the welding of the main structure, the cooling rate and restraint were high. The initial crack developed after welding. The crack of 75 mm long contained paint so must have existed before the structure was painted. The hydrophone mounting weld evidently was considered "non-structural". Thus even before launching the crack had propagated into brace "D-6" and then grew even longer as a result of fatigue loading of wind and waves and corrosion. Once it reached the critical length for the type of steel and the stress, brittle failure ensued.

Once brace "D-6" broke off the five other braces securing column "D" to the platform quickly broke off too. The structure was unstable in the water once column D was lost.

The original structure was statically indeterminate under gravity loading. However under lateral loading the compression stresses in the legs reduced. As the compression stresses approached zero the stability became critical (clearly no tensile capacity exists in each leg). However the structure could not maintain its stability even when one leg was lost: it was not designed to cope with the loss of one member. Thus the structure was **not** redundant.

The deck had been made top-heavy by the extra accommodation modules (raising the centre of gravity and making the structure less stable under lateral loads-the collapse was probably assisted by the wind) and that capsize was speeded by presence of open hatches and doorways.

Lessons

- An all-welded structure is completely continuous. Thus there is no such thing as a "minor weld". They should all be done carefully.

- Ensuring the structure is statically indeterminate is a necessary step but by itself is insufficient for real redundancy. To increase safety, design for the loss of one member.

- The hole in the brace "D-6" for the hydrophone was flame-cut making the edges brittle. Avoid flame-cutting.

- Preheat if necessary (say if using thick plates) to reduce effect of cooling.

- Avoid weld add-ons; if must use them ensure weld is properly stress relieved (e.g. by postheating or shot-peening) and carefully inspected.

Additional Information

A marine environment is corrosive and this can speed up the process of crack growth. If water is cold the steel material can be affected. Special steels are used to cope with the severe environment. These steels are supposed to retain their toughness even if temperatures drop to -40°C. In the case of the Alexander Kielland the steel was a weldable carbon/manganese steel of yield strength = 355 N/mm^2. Fatigue is a particular concern when the environment is also corrosive. Mild steel loses its fatigue limit when immersed in sea-water.

6.4 Steel Moment Resisting Frames, Northridge, California, USA, 1994.

[Shepherd & Frost (1995), Feld & Carper (1997), Taranath (1998)]

At 4.31 a.m. on Jan 17th 1994, there was an earthquake at Northridge, near Los Angeles, California. The earthquake had a Richter magnitude of about 6.8 and lasted 10 sec. It caused 61 deaths and economic damage over US$30 billion.

Steel Moment Resisting Frames are a very common seismic resisting system. Instead of the ductile performance of these structures, over 100 buildings suffered brittle fracture of their welded connections, compromising their ability to resist future shocks. It was a big surprise to the engineering community. There was an enormous loss of confidence in the system. The earthquake in Kobe, Japan in 1995 caused similar connection damage.

Much of the damage was concealed behind finishes. Very expensive retrofitting had to be carried out to repair the damage. Low, medium and high-rise buildings were affected. All buildings were designed and constructed in accordance with the latest earthquake resisting standards. The recorded accelerations were less than half those designed for.

Beam-Columns were connected using shear cleats and site welded flanges to give resistance to moments (a typical beam-column connection is shown in Figure 6.11). Cracking predominantly occurred in the site welded connection between the bottom flange of the beam and the column. Most of the buildings affected were less than 10 years old and had large moment connections.

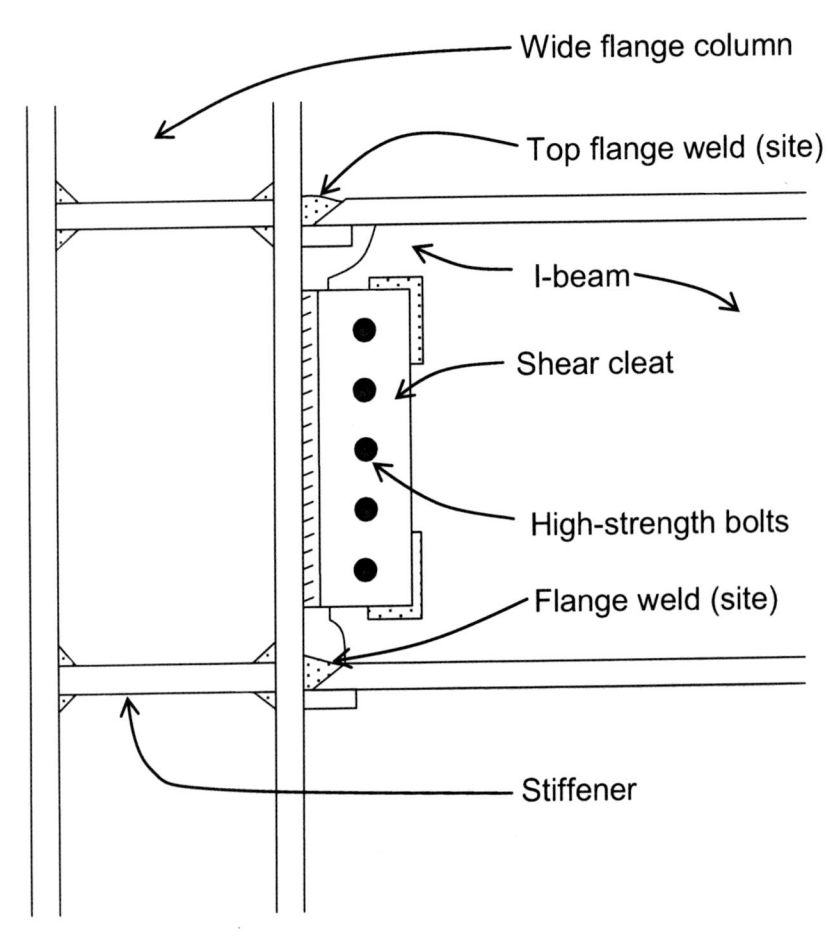

Figure 6.11: Common moment-resisting connection

To increase the speed of the welding, high welding temperatures, large welding beads, and less testing were becoming normal. This meant the resulting welds were less tough than before. The high strain rate imposed by the earthquake (0.01-0.1/s) exceeded the low toughness of the welds and the steel of the Heat Affected Zone (HAZ). To increase architectural freedom fewer, larger moment connections were being used.

The experiments upon which the designs were based were done in the 1970s using smaller moment connections, less typical of the larger connections of the 1990s. The welding procedures were those used in the 1970s.

Lessons

- Codes of practice, in particular those related to earthquakes, continue to evolve in a trial-and-error fashion. They are only as up to date as the experiments upon which they are based.

- As practice evolves it might move further and further from the original research and that research is no longer relevant. Practice is always ahead of research.

- It is recommended that designers use stiffners or haunches to force the plastic hinge to form within the beam, away from the column, and so do not rely on the ductility of the welded connection.

- The column is required to remain elastic under overstrength moments, typically $1.25 M_P$. (This is the "strong column, weak beam" philosophy, a philosophy relevant to all of structural engineering, not just earthquake-related). This approach encourages the plastic hinge to form in the beam rather than at the welded connection with the column, by strengthening the connection with stiffeners (shown in Figure 6.12) or by tapering the beam flange forming a reduced beam section (RBS).

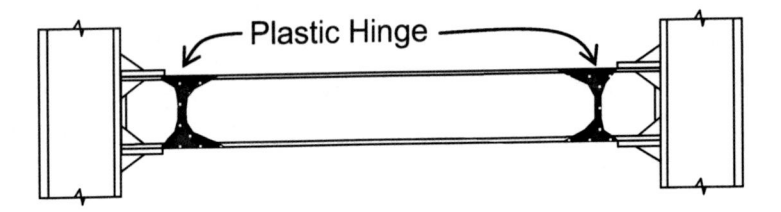

Figure 6.12: Encouraging the hinge to form within the beam rather than at the connection

Chapter 7: Failures during Construction

"Human beings, who are almost unique in having the ability to learn from the experience of others, are also remarkable for their apparent disinclination to do so."

Douglas Adams (Writer)

The following chapter considers the failure of structures during construction. As well as discussing several case studies, some general recommendations are also made.

Case Studies:

- Quebec Bridge, Canada, 1907.
- West Gate Bridge, Melbourne, 1970.
- Bailey's Crossroads, Virginia, 1973.
- Willow Island Cooling Tower, W. Virginia, 1978.
- Cocoa Beach Condominium, Florida, 1981.
- L'Ambiance Plaza Condominium, Connecticut, 1987.
- MRT Circle Line, Singapore, 2004.

- General Recommendations.

7.1 Quebec Bridge, Quebec, Canada, 1907.

[Francis (1989), Ferguson (1992), Petroski (1994), Schlager (1994), Petroski (1995), Shepherd & Frost (1995), Feld & Carper (1997), Jennings (2004), Delatte (2008)]

Quebec Bridge was a steel cantilever bridge being constructed over the St. Lawrence River. It was the world's largest of its type up to that date (the centre span was intended to be 549 m). The designer was Theodore Cooper who was then one of the most experienced bridge engineers in N. America (he was one of the inspection engineers for the Eads Bridge, St. Louis).

In the weeks before the collapse, as construction neared the centre of the span, bowing of struts became visible (up to 57 mm), rivets were shearing off and the structure was making creaking noises.

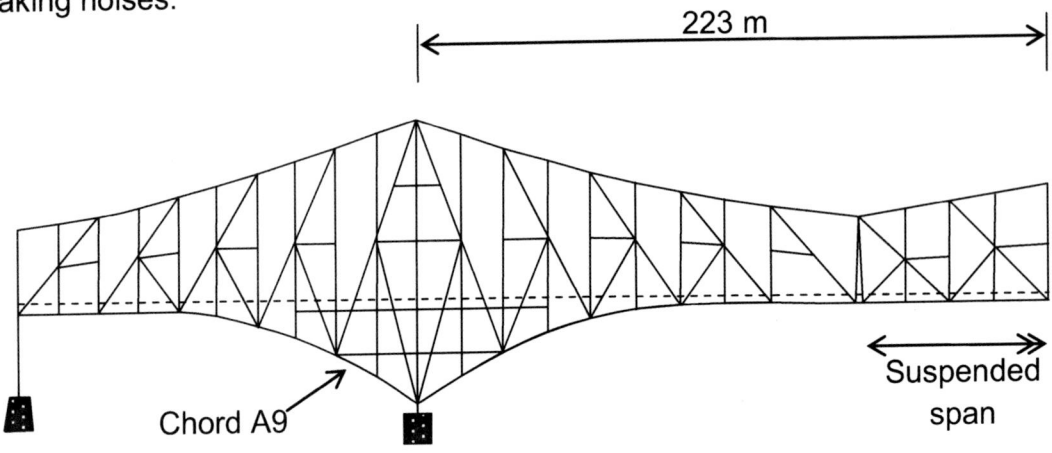

Figure 7.1: Configuration just before collapse

Collapse

On the morning of 29 August 1907 Cooper sent a telegram from New York to the site that work should be stopped pending investigation of the structural problems. Unfortunately it was not delivered in time. Later that day bridge collapsed killing 74 workers. Figure 7.1 shows the configuration just before collapse.

Investigation

The bridge collapsed straight down with no side sway. Workmanship was found to have been good.

Cooper, then aged 68, was unwell; he never visited the site during steel erection and only visited the fabrication shop three times. He worked as Consultant while the Contractor prepared the details. The Consultant accepted a fee much less than 1% of cost of bridge. It was inadequate to retain any design staff, so Cooper personally prepared the design and checked the Contractor's details. In addition, no experienced engineer was appointed to be on site during construction. Cooper ran the project from New York.

Cooper objected to any independent checks of his design. Thus the design was not checked. For example, in February 1906 when steel fabrication was underway, it was found that the dead weight was underestimated by 24%. It was estimated by Cooper that the stresses were increased by 10-25%. Nevertheless no change was made to the design: the increased stresses were simply accepted.

The trigger for failure was pair of struts "A9" in the anchor span. The struts were 17.4 m long and 1.4 m deep. They buckled due to inadequate lacing. Each strut consisted of four members laced together with light angles which had one rivet at each end of the lacing members.

The behaviour of such a built-up section is difficult to predict theoretically (especially since it can fail as a unit by torsional buckling), so even now we would rely on large-scale tests.

The lacing was designed on the basis of small-scale tests conducted 20 years before on struts with 1/30 cross-sectional area of "A9". Larger scale ($^1/_3$) experiments were done after the failure. They showed that a 50% increase in the size of the lacing angles, and the use of two rivets at their connections would increase the capacity of the strut by 12% *above* that which was estimated to be the failure stress of the unmodified "A9".

When completed, the bridge was to be longer than the earlier cantilever bridge over the Firth of Forth in Scotland (span 521 m built in 1890). However the compression members were clearly much smaller.

Lessons

- Accepting a low fee can lead to poor design.
- Small-scale tests may not be a reliable guide to the behaviour of the full-scale structure.
- The designer should always consider how other engineers have solved a similar problem (the Forth Bridge was a relevant example).
- Avoid situation where nobody on site can authorise work to be halted.
- Construction load cases can be onerous (the cantilever moment due to self-weight was about 10% larger during construction than in the final case, as the suspended span was finally made to be simply supported).

"We step up from the ordinary columns of ordinary construction, tried out in multiplied practice, to enormous heavy thick-plated pillars of steel, and we apply the same rules. Have we the confirmation of experiment as a warranty?"

Investigator of collapse.

"His [Cooper's] action was a grievous wrong to the engineering profession, as it tended to create the impression that responsible engineering service was of little account and could be had for next to nothing..."

Gustav Lindenthal (designer of Manhattan and Hell Gate bridges)

Aftermath

A replacement bridge was constructed. It had a similar form, i.e. cantilever bridge, but was substantially heavier. The construction method was changed: the complete suspended span was floated into position and lifted into place. During the first attempt at the lift a casting broke and the suspended span fell, resulting in 11 deaths. The completed bridge opened in 1918.

7.2   West Gate Bridge, Melbourne, Australia, 1970.

[Blockley (1980), Ross (1984), Francis (1989), Schlager (1994), Heyman (1999), Jennings (2004)]

It was one of the first welded steel box-girder bridges. It was to have a total length of 2,585 m.

The 25 approach spans, which were 67 m each, were prestressed concrete with the centre five spans being steel. The maximum span was to be 337 m and cable stayed.

The bridge was 37 m wide consisting of 6 lanes in each direction. The cross-section was a single three-celled box. The designers chose a box girder in order to maximize torsional stiffness (so avoiding Tacoma-type problems) and thin steel plates could be used (so avoiding the risk of brittle failure/welding difficulties, e.g., King's Street). So they limited the plate thickness to about 13 mm and in addition, no site welding was allowed. Connection between box girders segments was made by bolting using High Strength Friction Grip (HSFG) bolts.

Collapse

While under construction, a 112 m span steel box girder collapsed. It was simply supported during construction, later to be made continuous using concrete topping. The collapse caused 35 deaths and 15 injuries.

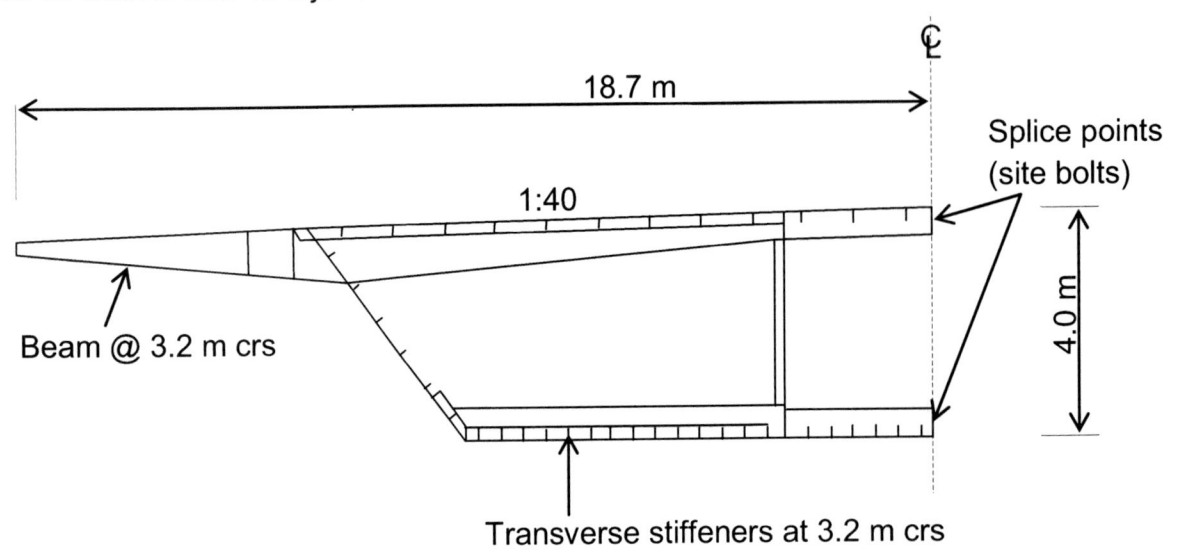

Figure 7.2: Half cross-section of bridge

Investigation

Originally two contractors were appointed, one for the steel part of the bridge and one for the concrete part. Construction began in 1968. The erection method for the steel portion was suggested by the contractor and approved by the designer: each span was constructed in two halves (Figure 7.2 shows one half) and bolted together along the centre-line. The advantage of this method would be that only half of the bridge would have to be jacked up the piers, which were about 54 m high, each time.

Fabrication began with a near-identical span of 112 m on the other side of the river. The full length spans were first assembled from prefabricated half-boxes 16 m long, on the ground. The two half-boxes making up full width of bridge were not symmetrical about a vertical axis because of road camber. Thus when lifted off the ground they bent away from their intended line as the neutral axis was not horizontal (Figure 7.3).

Once off the ground the full simply supported deadweight acted. The top plate of the box was thus in compression. Unfortunately the erection calculations (done by the original steelwork contractor and checked by the consultant) were over-simplified.

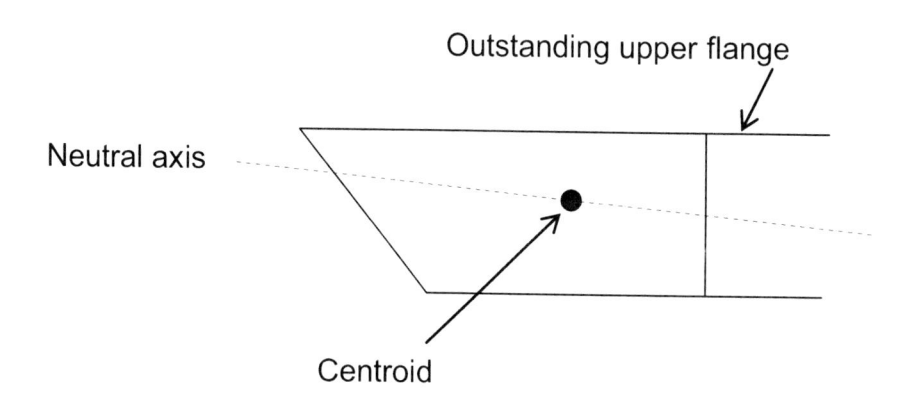

Figure 7.3: Neutral axis of half section not horizontal

Three important considerations make hand calculations very difficult. The simple "engineers beam theory" formula cannot be easily applied, i.e., $\sigma = My/I$ for the following reasons:

- As mentioned above, the half-boxes were unsymmetrical so bending was not about a horizontal axis;

- The presence of shear in the flanges cannot be ignored, (it is reasonable for a rolled I-section but not a box girder as the plates are thin.)

- Shear-lag effects must be properly accounted for (Figure 7.4).

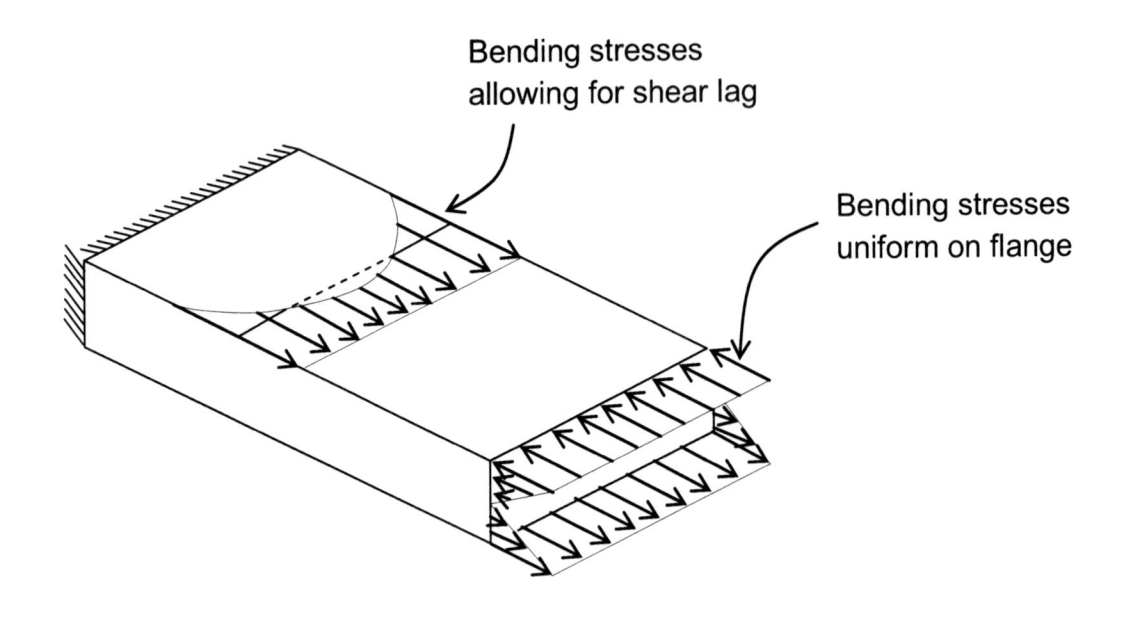

Figure 7.4: Shear lag effects

E.g., in the single-cell box illustrated: the flexural stresses are higher at the webs than at the centre. This is due to shear deformations of the flanges (i.e. flanges not close to webs).

These factors meant there were larger compressions in the flanges than would be expected. In addition, it is now better known that buckling resistance of plates is very sensitive to any pre-existing deformations or distortion due to temperature or welding. For example, if the thrust is eccentric by $t/6$ where t is the plate thickness, it means that the stresses are doubled. In this case there was inadequate stiffening provided to the top plates to cope with these compressions. They led to buckling of the top plate.

However, by removing some bolts the contractor somehow made the two halves fit. (The removal of bolts reduced the restraint to the top flange so was very risky). Thus the span was successfully erected. The team now felt they could deal with any buckling.

At this point, 1¹/₂ years into construction, the original steel erection was seven months behind schedule, mainly because of poor labour relations. So in an effort to "speed up" construction, the client appointed the <u>concrete</u> contractor as the steel erector for the remainder of the bridge. This contractor had no previous experience with steel erection.

Thus the erection of the next 112 m span began under the new contractor. Unfortunately the erection problems were much worse than the previous span. When the two halves were lifted onto the piers the difference in vertical alignment was 114 mm and the boxes were over 60 mm apart. Large buckles appeared as soon as the boxes were lifted.

As before, the contractor attempted to solve this misalignment in the air. However, this time the contractor added kentledge (8 tonne concrete blocks) to the span, as well as removing some bolts in an effort to make the halves fit and counteract the buckling. Complete collapse by hinging at mid-span followed soon afterwards.

The immediate cause of the collapse was the Contractor's removal of a number of bolts from a transverse splice in the upper plating near to mid-span in an attempt to join the two half- boxes. Things were made worse by the poor relations between parties in construction. That meant there was poor communications. The inquiry also blamed the Client for creating a climate of urgency and pressure which tended to lower morale.

Most of the blame, however, was put on the designers for:

- Adopting an untried, and so risky, method of erection.

- Using a safety factor for erection that was considered too low (1.31) given the state of knowledge of this structure type. This was the safety factor recommended by the relevant bridge design code for all bridges at that time (British Code BS153).

- Errors in the calculation not being noticed.

Lessons

- Thin plates can distort easily so the capacity of a box girder is frequently much less than the theoretical capacity ignoring this distortion.

- If relations between designer and contractor are poor there is an increased risk of the lack of communication causing structural collapse.

- The contractor must have appropriate experience.

- Some workers were killed as the bridge fell because their site huts were placed directly below the bridge.

- Avoid very wide single boxes. (The erection behaviour would have been much simpler if two boxes were used instead of one three celled box).

- Consider construction explicitly during the design process. (Design-and-Build contracts have this advantage).

- In general, bucking is a brittle failure and the factor of safety against it should be increased.

- Extra care is needed in preparing calculations if bucking is a risk, e.g., consider effects of support movement on the buckling length.

- During construction many structures are statically determinate and so more vulnerable to total collapse.

Quotes from investigation commission

"Every feature of the West Gate Bridge had been used before, but not necessarily in combination or on such a large scale."

"The engineers] failed to give a proper check to the safety of the erection proposals put forward by the original contractor."

"The margins of safety for the bridge were inadequate during erection; they would also have been inadequate in the service condition had the bridge been completed."

"Willingness to cope with imponderables is one thing that separates engineers from scientists, but it is imponderables that can provoke failures."

"The box girder bridge resembled a piece of aircraft structure, yet bridge builders are not able to test their structures in the same way that engineers in the aerospace industry do. Only after one full span and much of the second was built were defects observed."

Additional Information

In the previous June 1970 another box girder bridge, at Milford Haven in Wales, designed by the same firm of engineers, collapsed. Two other steel box girder bridges collapsed around that time (at Vienna in 1969, and Koblenz in 1971). In 1973 the UK Institution of Structural Engineers issued comprehensive guidelines that were eventually incorporated into the British bridge code BS5400 and thence into to the Eurocode EC3.

In 1978 the redesigned West Gate Bridge opened.

7.3 Baileys Crossroads Condominium, Virginia, USA, 1973

[Ross (1984), Hurd & Courtois (1986), Kaminetzky (1991), Schlager (1994), Shepherd & Frost (1995), Feld & Carper (1997), Delatte (2008)]

This case concerns a flat slab building that partially collapsed during construction. The tower was intended to be 26 storeys high. The floors consisted of flat plates of 200 mm thickness spanning up to 8.0 m and without any punching links at the columns. As is fairly common in apartment buildings the spans floor plate was irregular. The collapsed zone is indicated in Figure 7.5.

Collapse

The collapse took place as construction of the third identical 26 storey tower was nearing completion: the contractor was placing wet concrete on level 24. For some reason the contractor decided to remove the props under the level 23 slab (specifications required that at least 3 levels of propping remain in place). This meant the entire weight of the wet concrete on level 24 as well as its own weight was taken by the level 23 slab. Unfortunately, punching failure of the concrete of the level 23 slab took place and propagated vertically as the falling slabs collapsed the levels below. Eventually the failure stopped at the ground level cutting the building in two. The gap formed was 20 m wide. Fourteen workers were killed in the collapse.

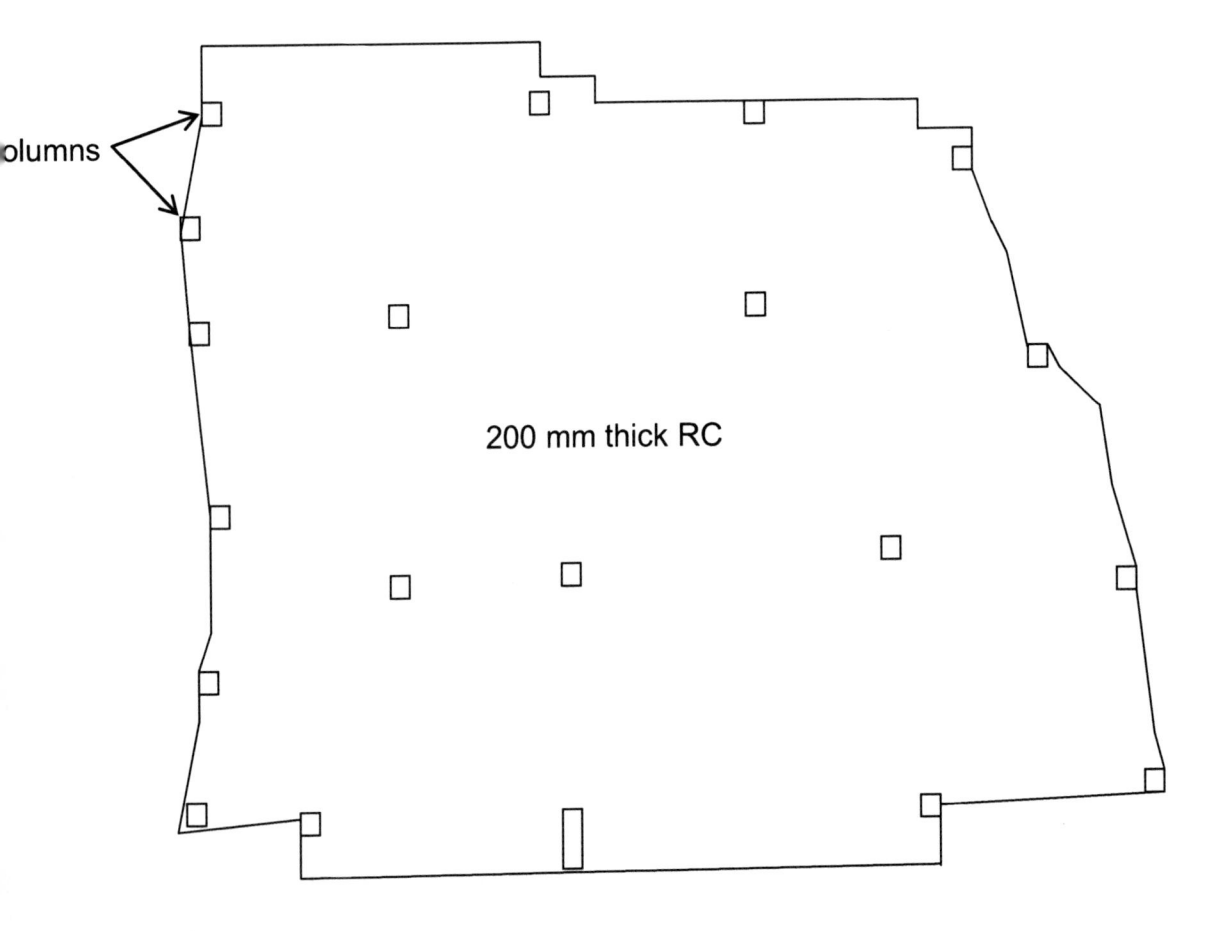

Figure 7.5: Irregular floor plate spans

Investigation

The code used to design the structure was ACI-318: 1970. Unfortunately only versions of that code issued after 1989 required bottom bars to be placed through the column cage to help prevent progressive collapse. The ultimate punching stresses were within the limits of the ACI code based on concrete of 28 day strength resisting design loads.

The contractor's practice was to cast 1 floor every 7 days. In the days just before the collapse the cycle accelerated to one floor every 3 days. In the cold weather the strength gain of the concrete was especially slow. Cores taken from the upper slabs indicated strengths of only 5-10 MPa.

The contractor was obviously at fault and was blamed by the court. While the engineer's contract specifically required no site inspection, the engineer was nevertheless blamed for not insuring inspection was *done by others*. As the engineer and architect had large insurance policies, it was they who were sued by an injured construction worker for over $500,000.

Lessons

- Bottom bars should always be used to prevent progressive collapse.

- Be especially careful of punching of low additional dead + low live load structures during construction and low live load structures during service.

- Ensure the contractor takes proper matched-cured samples (samples cured beside the slab they represent) or conducts test on the actual structure, to estimate *in situ* strengths, before de-propping begins.

- Include all gravity loadings that have been used for the final design of the floor on the drawings.

Additional Information

Construction loads may be larger than final design loads, particularly of condominium buildings and other structures of low live load design capacity.

For example, if D is the self-weight of a typical floor and two levels of propping are used, the floor slab beneath these props is subject to a maximum load of about $2.25D$ during the construction of a multi-storey building (see Figure 7.6).

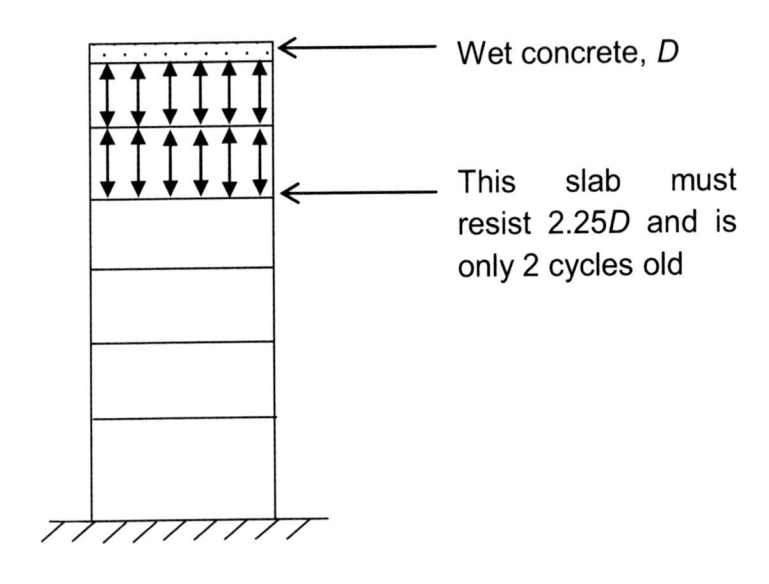

Figure 7.6: Construction loads

This ultimate strength must be available to avoid cracking, additional deflection, or even collapse (Hurd & Courtois 1986).

In addition, the early removal of props during construction is known to have caused large deflections in slabs in service. According to Ratay (2010), re-insertion of removed props (say after formwork is stripped) does little to reduce long-term deflections.

7.4 Willow Island Cooling Tower, West Virginia, USA, 1978.

[Ross (1984), Kaminetzky (1991), Levy & Salvadori (1992), Schlager (1994), Feld & Carper (1997), Delatte (2008)]

The construction of a 131 m high hyperbolic cooling tower was completed up to about 52 m. The walls were 200 mm thick. The system of jump form scaffolds in use (Figure 7.7) had been successfully used for the construction of 36 other towers. The scaffolds were anchored to the concrete walls using expansion bolts. The temperatures were low (17°C during day, 0°C at night).

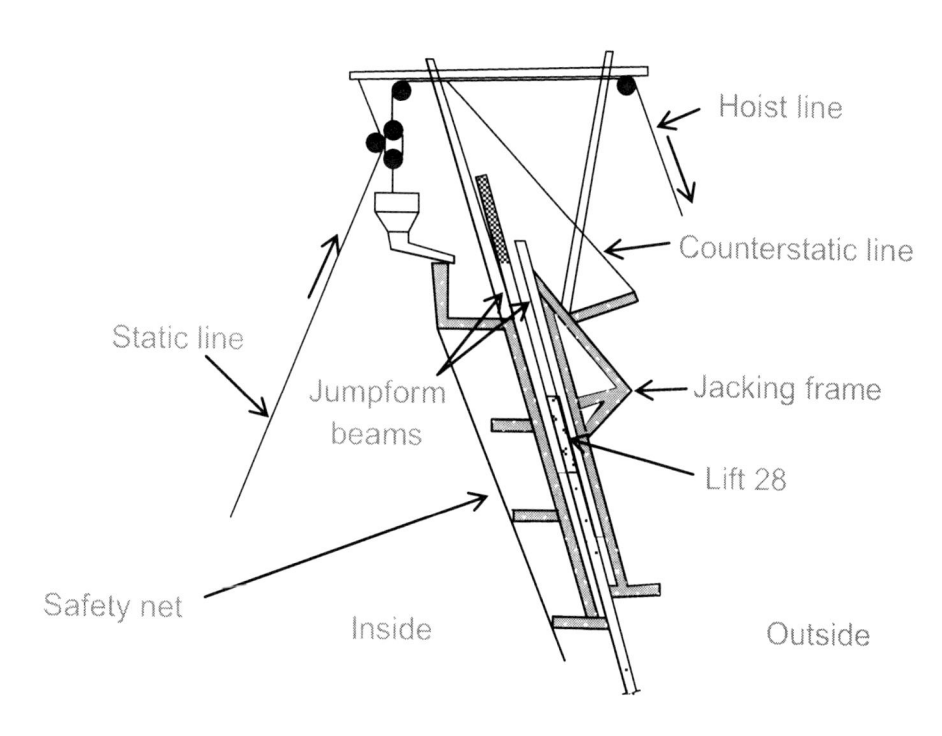

Figure 7.7: Jump form system

Collapse

A ring of formwork was suspended near the top of, and inside, the cooling tower. The anchors holding the scaffolding pulled out of the concrete and the workers fell. Several of the workers had successfully jumped to an adjacent safety net, but still lost their lives as the collapsing portion dragged the safety net with it, killing 51 workers.

Investigation

Aluminium beams supporting scaffolding were anchored to a 1.5 m lift of concrete cast 24 hours earlier (lift 28), using two bolts each. Failure occurred when a new batch of concrete (intended for lift 29) was being hoisted from the ground. The anchors pulled out and workers fell 52 m.

The forms were usually raised hydraulically on a one-day cycle. However, the concrete had not sufficiently matured to support the formwork, scaffolding, workers and concrete hoist. The concrete was high strength (f_{cyl} = 38 MPa). No testing of the actual capacity of embedments by contractor took place so the low strength of the concrete was not detected.

It was later estimated (using the actual mix supplied and curing at 4.4°C) that the compressive strength of the 20-hour-old concrete was only 1.5 N/mm^2 while a minimum of about 6.9 N/mm^2 was needed using the jump form system. At the time of failure the concrete bucket was in transit (using a cable) and 18 m or so below the point of collapse.

In addition, it is likely the scaffolding was not properly secured to the concrete. The lower bolts of the formwork system may have been left out. The entire formwork system was connected together, so once part of it fell the whole fell.

Lessons

- The strength of an embedded anchor depends greatly on the *actual* strength of the concrete *not* that of cured samples.

- Find out the actual strength of the concrete using pull-out tests prior to removal of forms and shoring.

- It is often uneconomical to provide the formwork system with real redundancy. For such temporary structures an appropriate approach is to use joints to *isolate* sections so that if one part falls the rest is not pulled down too.

Additional Information

In 1981 a similar jump-form construction system was being used to construct another cooling tower. It collapsed; however "only" two workers died. A lesson learned at Willow Island was applied, i.e., that all the forms would fail progressively if they were tied together. The forms were assembled separately, in sixty sections.

Note: How can tying the structure together be "bad"? Consider the following illustration (Figure 7.8). The effect of tying the structure together is to increase the connection loads in the event of an accident. Thus the effect of the accident can more easily spread.

Tied

Now consider loss of a support

Increased reaction

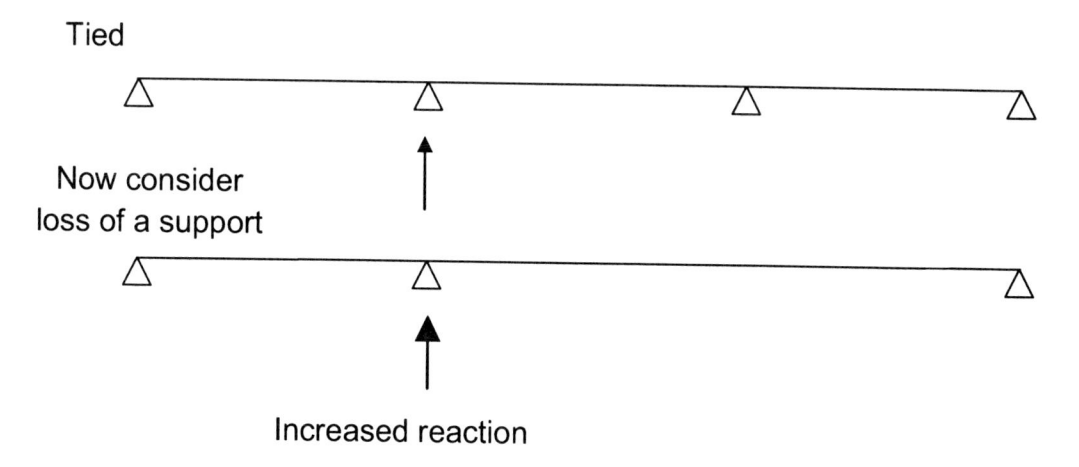

Figure 7.8: Tying structure together can lead to increased connection loads

7.5 Cocoa Beach Condominium, Florida, 1981.

[Kaminetzky (1991), Schlager (1994), Feld & Carper (1997), Delatte (2008)]

Cocoa Beach condominium was a five-storey RC structure 74 m long and 18 m wide. Its floors were 200 mm thick flat plates which spanned up to 8.4 m. They were constructed using flying forms.

Collapse

The fifth floor was being constructed when the formwork between level one and two was removed. The structure totally collapsed, killing 11 workers and injuring 23.

Investigation

Workers were finishing part of the 5th storey slab at the time of the collapse. Those that survived reported hearing a loud crack which sounded like wood splitting and then seeing the slabs "just sliding right down around and over the columns". Clearly they were describing a punching failure.

On looking at the design drawings it was clear there was a glaring mistake: the slab was much too thin: the ACI recommendation for the span and column size was 280 mm. The designers had been made aware of severe cracking and large deflections (up to 44 mm) noticed several days before the collapse (Figure 7.9). They responded that everything was "okay" with the design.

The engineers were retired NASA aeronautical engineers. They surrendered their licenses after the collapse. Punching shear calculations had never been performed. No deflection checks were done either.

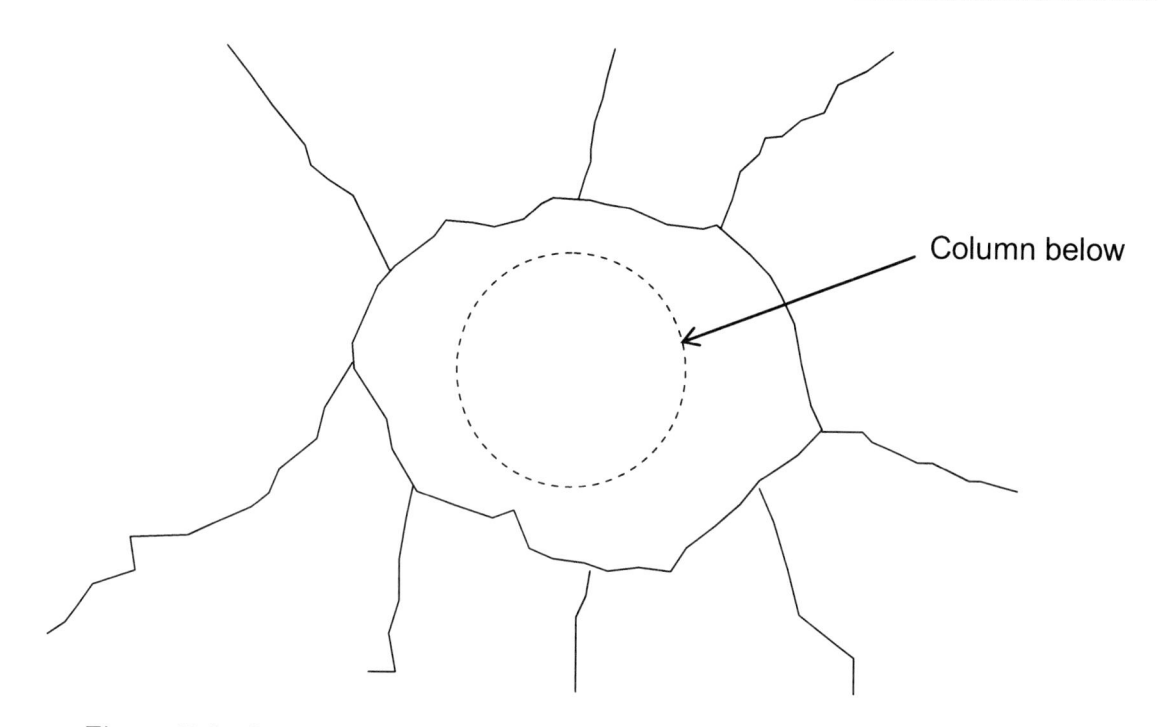

Figure 7.9: Cracking visible at the top of the roof slab directly above a column indicating that punching failure is imminent

The construction was poor too: reinforcing bars were placed in the top of the slab such that the effective depth was only 135 mm. The column steel was quite congested especially at splices, making concreting difficult. *Laboratory*-cured concrete samples were used to determine when slab stripping could take place rather than *site*-cured. Concrete maturity was also unfavourable: the collapse happened in March when temperatures were low.

The reason the structure lasted until the construction of the roof before collapsing was simply that all floors were completely propped until the construction of the roof was almost complete. Only when the props began to be removed were the floors required to carry significant load.

Lessons

- Ensure the design is prepared by qualified people.
- Use bottom bars passing through the column cage to prevent progressive collapse.
- Designs for projects should be subject to peer-review.
- Make sure the columns are big enough to avoid excessive congestion of reinforcement.
- Follow up on any on-site observation of trouble.

"If you think hiring a professional is expensive, try using an amateur!"

(Red Adair)

7.6 L'Ambiance Plaza condominium, Connecticut, USA, 1987.

[Kaminetzky (1991), Levy & Salvadori (1992), Schlager (1994), Shepherd & Frost (1995), Feld & Carper (1997), Delatte (2008)]

L'Ambiance Plaza was intended to be a sixteen storey apartment structure. There were two wings (East and West). Steel columns were first erected in sections. Flat plate slabs (180 mm) were post-tensioned using unbonded tendons. The building was constructed using the lift-slab technique.

Collapse

All floors had been cast and most slabs were in their permanent positions. No final welding had yet been done on upper floors. It was during the jacking operation that a loud metallic noise was heard, the level nine floor slab cracked up, and the entire building collapsed, killing 28 workers.

Investigation

The contractor involved in this collapse had previously built over 800 structures in the USA using this method. The lift-slab technique was invented in 1948 and had a good safety record before this collapse. It is considered safer than convention construction as less work has to be done at a height.

The entire building fell in only five seconds, 2.5 seconds longer than it would take and object to free fall from roof height.

The case was settled shortly after investigations began. Thus the investigation was never completed and so the precise cause remains a matter of debate. Here is one theory that many accept. The others are presented in Delatte (2008).

Most investigators agreed the collapse started at level 9 of the West wing. It was likely that the trigger was a failure of the lifting system (Figure 7.8). Laboratory experiments later confirmed that the deformation of the lifting collar angle was excessive. This resulted in the jack rod slipping off the lifting angle. The metallic sound heard at the beginning of the collapse may have been this rod striking the column.

As one investigator noted, lift-slab techniques and details have evolved over the years to the point that some of the fail-safe details and safeguards responsible for the good safety record may have been "optimized" out of the process.

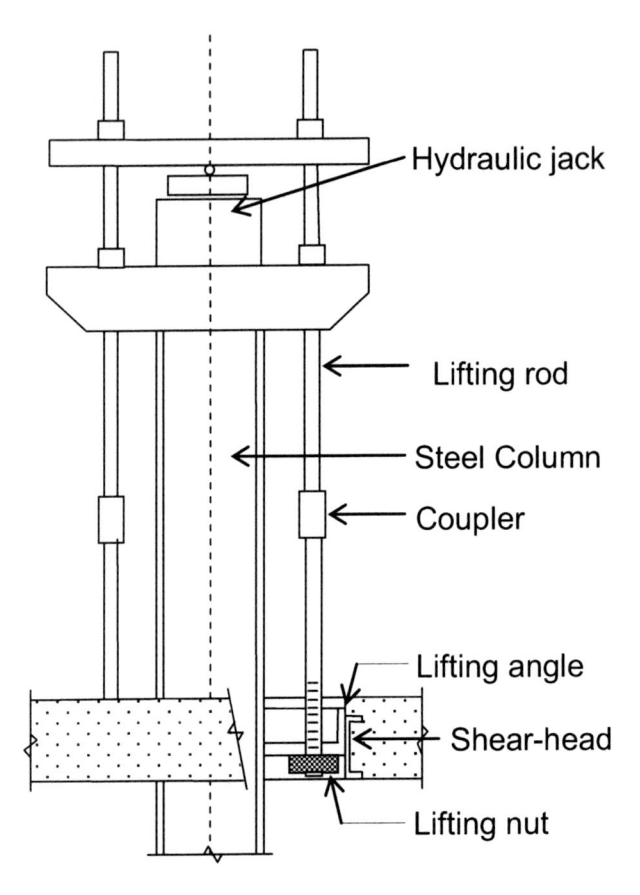

Figure 7.10: Lifting Mechanism

Regardless of the role in this failure, the temporary stability of the structure is crucial; temporary slab-column connections were unreliable and could not ensure temporary frame stability. There was no temporary bracing or timber propping which were standard features in the years before. The permanent bracing was by shear walls, however their construction lagged too far behind. They should really be at most 2 storeys behind.

The project delivery system was particularly complicated and fragmented. Responsibility for the ultimate structural safety of the building was confused by unclear relationships among the design consultant, the lift-slab contractor, and the shear-head designer.

Lessons

- The lift-slab technique is particularly vulnerable before final welding and construction of the shear walls. Temporary lateral bracing must be provided.

- The project delivery system should be as simple as possible so that lines of responsibility are clear.

- Be especially careful of structures designed for low live loads during construction.

Aftermath

The ACI-318 code was revised in 1989 to include a requirement for the provision of additional "structural integrity" reinforcement when unbonded post-tensioning is used.

For lift-slab structures this requires that conventional unstressed reinforcement is present in the bottom of the column strips.

$30 million was paid to compensate victims' families. In addition they received part-ownership of the new replacement structure for L'Ambiance Plaza.

7.7 MRT temporary works, Singapore, 2004.

[Magnus (2005)]

This failure occurred to the temporary works of a deep excavation for a new line of Singapore's Mass Rapid Transit (MRT) system. The construction was for a new cut-and-cover tunnel. The soil consisted of much reclaimed soil and marine clay. The concrete diaphragm walls were strutted by steel sections and slabs of jet grout. The diaphragm walls were not part of permanent works. The walls and strutting were designed by the contractor. The strutting is shown on plan in Figure 7.11.

Collapse

In the area designated "M3" the excavation was over 30 m deep, while in the "M2" area, it was over 25 m. The workers were placing 10th level struts.

At 8:45 am: workers heard noises from the steelwork in the "M3" area ("thung" sounds: probably buckling). At 9 am: a worker saw that a channel stiffener and waler flange had buckled. At 2 pm: workers attempted to use lean concrete to stop the steelwork noises by placing it in the tops of walers and on the floor of the excavation.

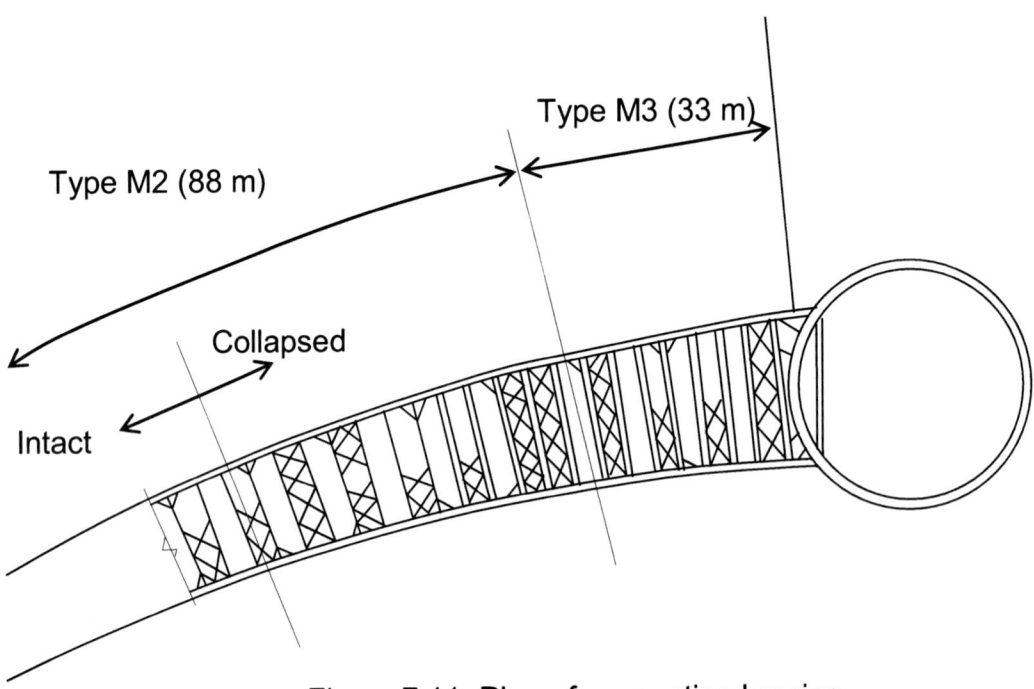

Figure 7.11: Plan of excavation bracing

The workers reported that the steel was making sounds like gunshots. Freshly placed concrete fell out as the walers had distorted so much. The instruments showed that loads in the struts on 9th level were falling, and those on 8th rising. At 3:30 pm: the walls caved-in from both sides. Diaphragm wall panels collapsed over length of about 80 m (all of M3 and much of M2), killing 4 workers.

Investigation

There were many warnings of impending failure: excessive wall deflections and inclinometer readings, water leaks, noises, visible distortion of steel.

The actual collapse began with lowest brace (level 9 was lowest active brace). It progressed vertically upwards before progressing horizontally, probably via the steelwork bracing (much of which was continuous along the length of the excavation).

A typical diaphragm wall section was 800 mm thick; each wall panel was 6,000 mm wide. The typical strut and waler were steel I-sections 400 mm overall. The typical strut spacing was 4 m. The collapsed area had 10 levels of steel strutting and in addition there were 2 levels of jet-grout slab. The upper level of jet-grout was "sacrificial". The lower jet-grout slab's purpose was to reduce water seepage. It was placed at the bottom of the excavation.

Pre-tender soil investigation data was presented in a report to tenderers. The soil data was "representative" or "moderately conservative" (requiring minimum Factor of Safety = 1.4 according to client's specification). However it was interpreted as "worst credible" by contractor. Thus the factor of safety used for design was only 1.2.

The excavation support was designed using "Plaxis" a 'general purpose' geotechnical finite element program. It was used to assess strut forces and wall moments. The design was to British codes BS5950 and BS8110 for the steelwork bracing and concrete wall respectively.

Conventionally, the web of an I-section is stiffened using a plate. However, in this case the contractor chose to use a C-channel section attached to both flanges instead. The use of a C-channel meant that ultimate failure of the struts was brittle. Conventional bracing fails in a more ductile way.

In addition, there were two errors in the steelwork calculation for the level 9 strutting: splays were assumed to exist where there were none; and the "stiff bearing length" used was incorrect.

The Contractor's Plaxis analysis used "effective stress" soil parameters. However, if "total stress" soil parameters were instead used, (as recommended by 3 independent experts before the failure took place), bending moments in the wall and wall deflections increased by 100% and strut forces by up to 10%.

Summary of mistakes:

- Under-design of the steel strutting (resulting in it having only about 50% of the assumed strength).

- The factor of safety of the design was only 1.2.

- Incorrect soil model used (substantially underestimating wall moments and deflection, and underestimating strut forces by about 10%).

Lessons

- Temporary works should receive the same factor of safety as permanent works. (The Contractor used 1.2 and "moderately conservative" soil parameters apparently as the works were not permanent).

- For temporary works, progressive collapse resistance is best provided by **isolating** sections of the construction so that collapse cannot progress horizontally as it did in this case. Alternatively design the strutting so that if there is a loss of one strut then the load can be redistributed to others.

- The user of the software should have experience in its correct use: Plaxis could model the system adequately, but the wrong soil model was chosen.

"The increase in continuity that often accompanies the provision of alternative load paths may, in certain cases, promote rather than prevent failure progression."

Uwe Starossek (author of 'Progressive Collapse of Structures' 2009)

Aftermath

The Contractor was blamed for the collapse. The Professional Engineer (PE) who endorsed temporary works was blamed by the inquiry.

The Client now requires all temporary works for deep excavations to be independently reviewed.

General Recommendations

The standard practice is to accept a higher risk for temporary support structures because they are "temporary". In my view this is illogical, since structures are often at their most vulnerable during construction, e.g., statically determinate and/or without critical bracing of permanent structure. Thus the same, or perhaps even greater, factors of safety should be used.

In situ concrete

A frequent cause of failure of concrete structures during construction is failure of the supports for wet concrete. Formwork for slabs is a potentially unstable configuration (a heavy mass-the liquid concrete-at the top of a tall structure-the formwork). This is vulnerable to a P-Delta failure. Formwork is rarely designed for redundancy and often improperly braced. In some cases the bracing is totally absent. There may be inadequate foundations for props or there may be initial lack of plumb of props. Vibration of concrete and movement of construction equipment can impart sufficient lateral load to cause stability failures. In addition, stockpiling of building materials on completed floors may impose a load the floor or its supports cannot resist.

Good practices include construction of columns at least one day ahead of slabs to increase the stability of formwork when the slab is constructed, and using deliberately introduced breaks in the temporary works to ensure any collapse does not spread.

Steelwork/precast concrete/timber

Construction workers are not qualified to determine if construction needs temporary bracing. Unfortunately they are unlikely to know when to seek help either. Their fatal attitude is apparently *"if we work fast enough we won't have to brace it"*.

The Eurocode EC3 requires that the contractor is informed **by the designer** if temporary bracing is required.

Avoid "late elements", say masonry walls, as permanent bracing to "early elements", say steel/precast concrete structure. These are known as "non-self supporting structures" (in the sense that something vital is missing if only say the steel frame is constructed). Otherwise ensure temporary bracing is provided. In steelwork, begin erection with the construction of the braced bay.

Chapter 8: Fire, Explosion, and Impact

"Structural Engineering is the art and science of moulding materials we do not fully understand, into shapes we cannot precisely analyze; to resist forces we cannot accurately predict, all in such a way that society at large is given no reason to suspect the extent of our ignorance"

James Amrhein (director, Masonry Institute of America) quoted in Feld and Carper (1997).

In these case studies the performance of bridges and buildings under extreme loadings is analyzed. The case studies are:

- Ronan Point, London, 1968.
- Sunshine Skyway, Florida, USA, 1980.
- Abbeystead, Lancashire, UK, 1984.
- Alfred Murrah Building, Oklahoma, 1995.
- World Trade Center 1&2, New York, 2001.
- The Pentagon, Virginia, 2001.
- World Trade Centre 7, New York, 2001.

8.1   Ronan Point, London, UK, 1968.

[Francis (1989), Kaminetzky (1991), Levy & Salvadori (1992), Schlager (1994), Petroski (1994), Elliot (1995), Shepherd & Frost (1995), Feld & Carper (1997), Wearne (2000), Kletz (2001), MacLeod (2005), Delatte (2008)]

This case concerns the failure at a residential tower block known as Ronan Point. After the destruction of World War 2, the UK needed large quantities of new buildings quickly.

Precast concrete seemed an obvious choice. Various large-panel systems of so-called industrialized buildings were used. The Nilsen-Larsen system of large panel construction was used for the new 22-storey blocks at Ronan Point. Storey height precast reinforced concrete panels were comprised the walls and the floors consisted of precast units spanning between them. The joints were to be grouted together on site; however no reinforcement was specified to provide a sound connection in this case.

Collapse

One morning, shortly after the building had been completed and occupied, the owner of an apartment on level 18, a Mrs. Hodge, lit the stove in her kitchen to make tea (Figure 8.1). She did not know it but the piped gas system installed in the building was leaking and the rooms of her apartment were full of gas. There was an explosion. The South and East walls of the living room were blown out. The walls above them thus lost their support, as did the floors they supported. The impact of the debris on levels below caused them to collapse too. Thus the collapse was progressive. An entire corner of the building collapsed to the lowest level. Mrs. Hodge was deafened by the explosion and received severe burns but was otherwise unharmed. However, four people were killed on the floors below.

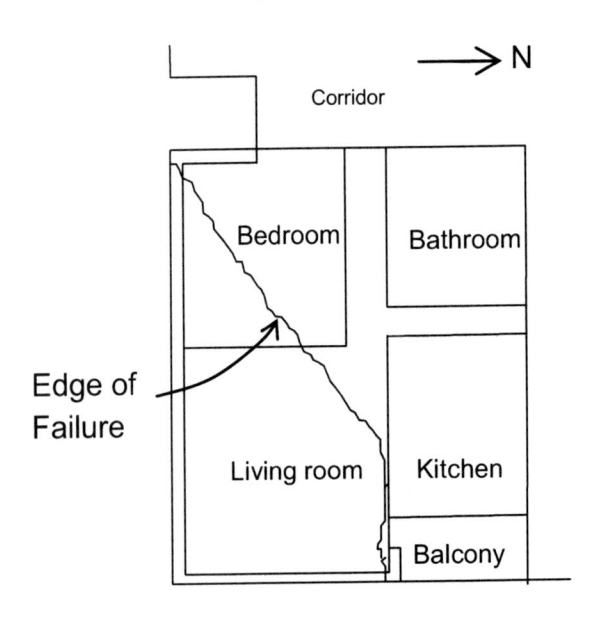

Figure 8.1: Mrs. Hodge's unit at Ronan Point

Investigation

The inquiry found that, as essentially only friction held the wall units in place, a pressure of only about 21 kN/m² was required to dislodge them at level 18. The code used for the design of Ronan Point was the British code *CP114 Code of Practice for the Structural Design of Reinforced Concrete Buildings*. However, this code was intended for conventional *in situ* structures made up of beams, slabs and columns. This structure had none of those *in situ* elements. Thus the code did not address connection design and the fact that it is essential to tie the precast units together to prevent "a house of cards" type of behaviour. One method of tying the units together is shown below (Figure 8.2).

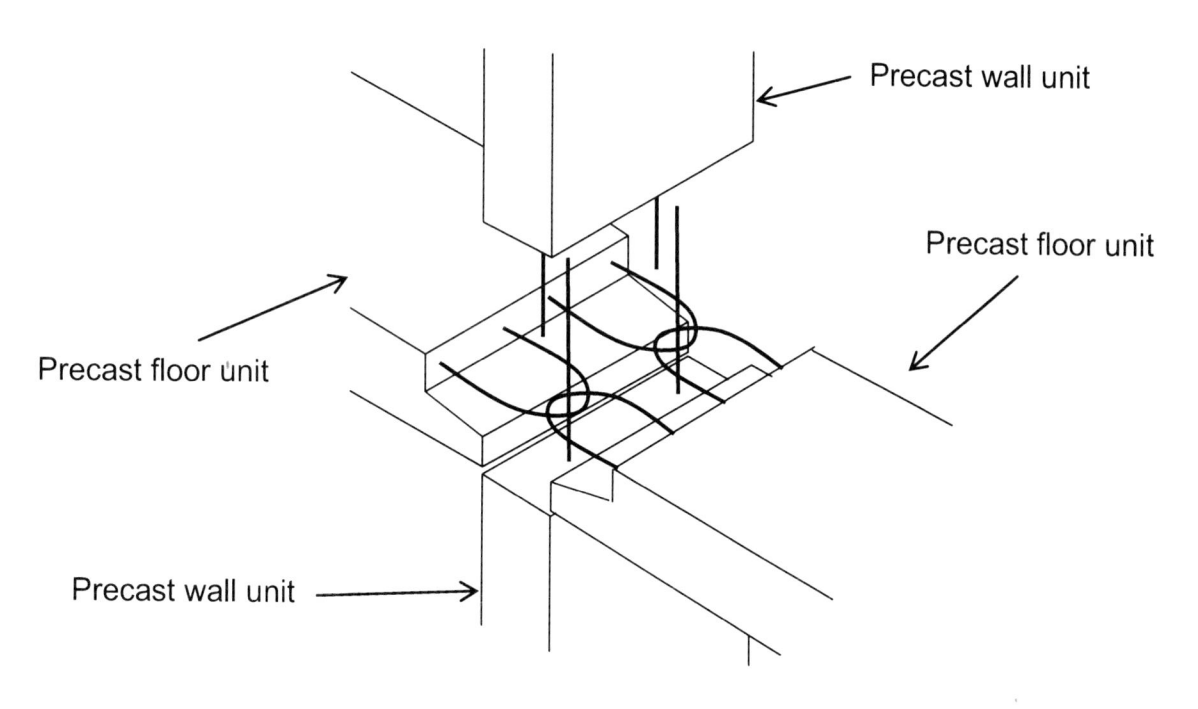

Figure 8.2: Tying of the precast units together using looped reinforcement

Later British codes were revised to make the consideration of progressive collapse mandatory.

Lessons

- Ensure the structure is adequately tied together to resist internal explosions.

- Ensure the scope of the code of practice used covers the design of the type of structure envisaged.

- Lightweight non-load bearing cladding is better than heavy load bearing cladding as it can be quickly blown off so the explosion is vented and so less severe.

- As large quantities of gas are needed for an explosion of an air/gas mixture, cylinder gas is "safer" than piped gas (as any gas leak can remain undetected for a long time).

- The probability of an explosion is low, but not zero, so we must lessen its impact if it does happen.

8.2 Sunshine Skyway Bridge, Florida, USA, 1980.

[Blackhall (1989), Schlager (1994), Feld & Carper (1997), Perrow (1999), Starossek (2009)]

The bridge opened in 1953 over busy shipping lanes. It was a steel cantilever truss supported on concrete piers. The structure was essentially statically determinate. There were two independent bridges: one for northbound road traffic and one for southbound traffic. Before the failure it was recorded that ships had struck the bridge piers at least eight times before 1980.

Collapse

The ninth accident proved fatal. It happened at 7:30 am on May 9, 1980. There was bad weather at time; there was a rainstorm with strong winds and there was poor light. A freighter weighing over 20,000 tonnes struck the pier of the southbound bridge, itself a 12,000 tonne structure. Three spans (427 m) of the southbound bridge fell. Thirty three people were killed. The structural arrangement was as shown in Figure 8.3.

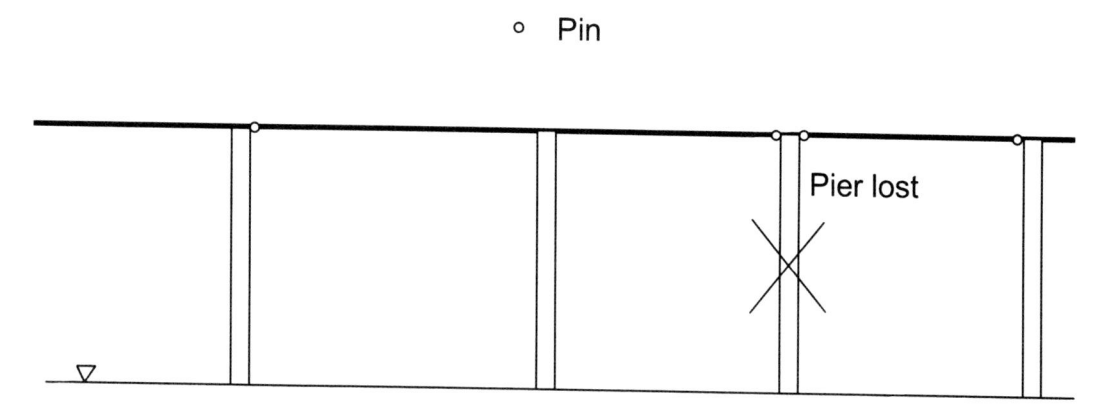

Figure 8.3: Structural configuration

Lessons

- Avoid statically determinate bridges, especially if no consideration of the collapse behavior has been made.

- If possible, avoid placing piers near the shipping lanes, by increasing the span.

- Otherwise protect piers adequately using huge concrete bases known as "dolphins".

- Alternatively design bridge so that the loss of a pier in possible. In the case of a bridge, catenary action is likely to be impractical. So some (limited) collapse is inevitable.

For example, the approach taken for the collapse design of the Confederation Bridge is given in (Starossek 2009). This 12.9 km bridge has 43 main spans of 250 m each. See Figure 8.4. The design considered the loss of one pier. If the span was doubled, relying on catenary action over 500 m span was impractical. It was found to be more cost effective to design bridge for loss of pier accident than to try and protect the piers. In the final configuration of the bridge, this resulted in statically determinate drop-in spans limiting the extent of the collapse to about 600 m or so. Thus this is an exception to the "avoid statically determinate structures" recommendation, as sometimes, particularly in bridges, such "fuses" can be the only way to achieve progressive collapse resistance.

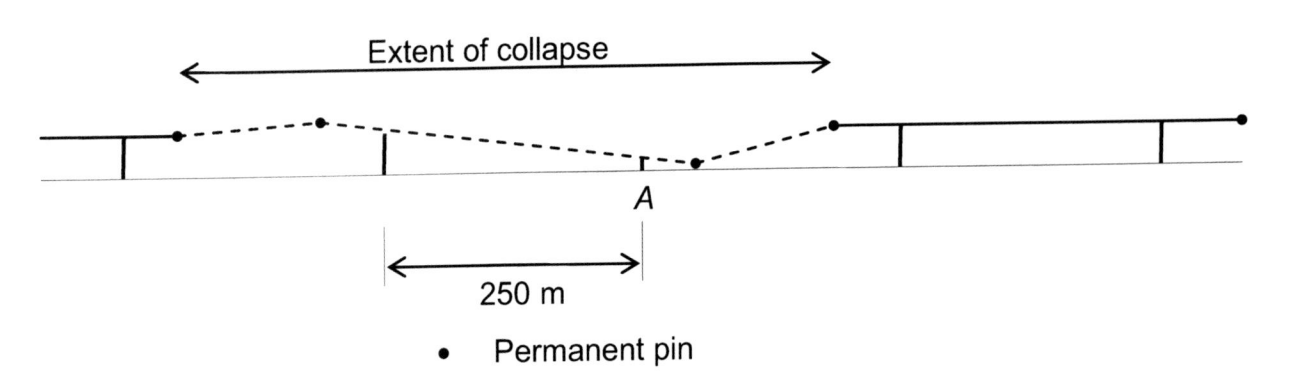

Figure 8.4: Collapsed configuration due to loss of pier at A

8.2 Abbeystead Valve House, Lancashire, UK, 1984

[Wearne (2000), Kletz (2001)]

This underground structure was in rural Lancashire. New towns in the area required water from the nearby Wyre River. However, the water in Wyre River was insufficient. New reservoirs were not allowed. Therefore the water was pumped, along 12 km of pipeline and tunnel, from the river Lune to the river Wyre where it was extracted. The structure was constructed in 1979. The arrangement is shown in Figure 8.5.

The new water system was controversial – in the early years of its operation it possibly caused flooding of land around river Wyre so it was not popular with the local residents. Thus as a public relations exercise a tour of Abbeystead Valve House was organized for 23 May 1984.

Collapse

A group of 44 people entered the main part of the valve house. The pumps were turned on. This was the first time they had been operated for 17 days. There was an explosion shortly after, which killed 16 people.

Investigation

The 2.7 m diameter tunnel linking the two rivers was lined with concrete. It was not designed to be watertight. This was to ensure that the tunnel would always be full of water, either water from the river Lune or groundwater. The inflow through the tunnel walls was estimated to be 264,000 gallons per day. However, clear evidence was found that the tunnel was not full of water for long periods of time, e.g., tide-marks on walls, stalactites growing. The washout valve near the Valve House had been left open by mistake allowing the water from the tunnel out. The tunnel was not ventilated. Numerous old books and maps in the local library mentioned that the area around Abbeystead was used as a coal mine hundreds of years in past. Coal is associated with the presence of methane.

During construction, traces of flammable gas were discovered on three occasions, by their smell or by the bubbling of water leaking into the tunnel.

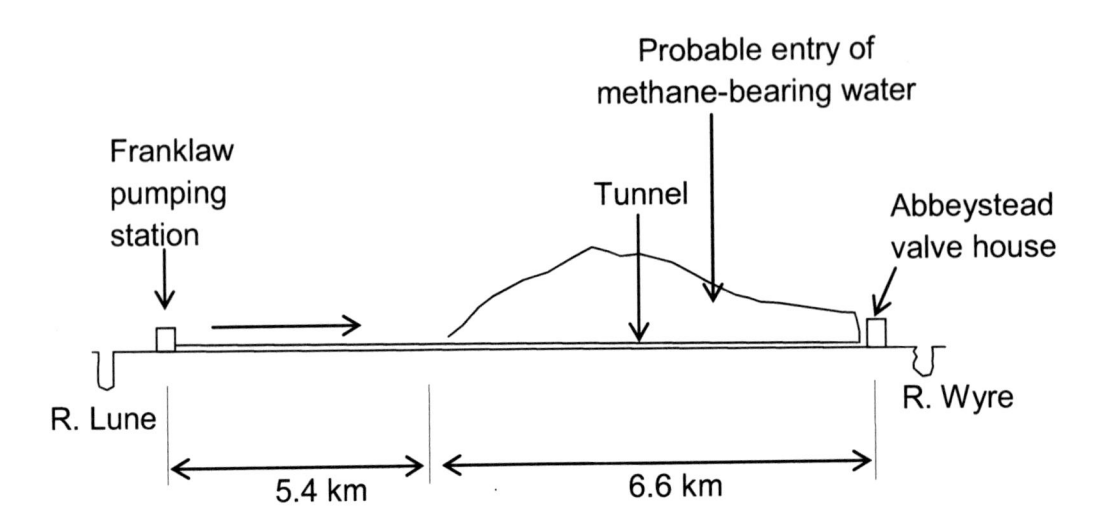

Figure 8.5: Cross-section showing layout of connection between rivers.

Lessons

- Essential to understand the geology of the area thoroughly if underground structures are planned.

- It must be carefully checked that an underground structure remains full of water at all times to reduce the chance of gas accumulating (methane is not water-soluble).

- A long tunnel must be ventilated.

Aftermath

Authorities declined to prosecute. A civil action was taken by relatives of the dead and injured. In 1989 the Water Authority, the designers and the Contractor were found negligent and fined £2.2 million.

8.4 Alfred Murrah Building, Oklahoma, USA, 1995

[Wearne (2000), Campbell (2001), Delatte (2008)]

The building was nine-storeys high and housed offices of the US Federal Government. It was an *in situ* RC one-way beam and slab building with glass cladding. The structure was designed to resist gravity loads and winds. It was constructed in 1974. The lateral loads were taken by RC shear walls.

Collapse

Shortly after 9 am on April 29, 1995 a bomb, consisting of Ammonium Nitrate and Fuel oil (estimated to be equivalent to 1814 kg of TNT) packed into a van, exploded just in front of the north face of the building. Nearly half the structure collapsed, killing 168 people and injuring over 500. The blast caused a crater nearly 8.5 m in diameter. See Figure 8.6.

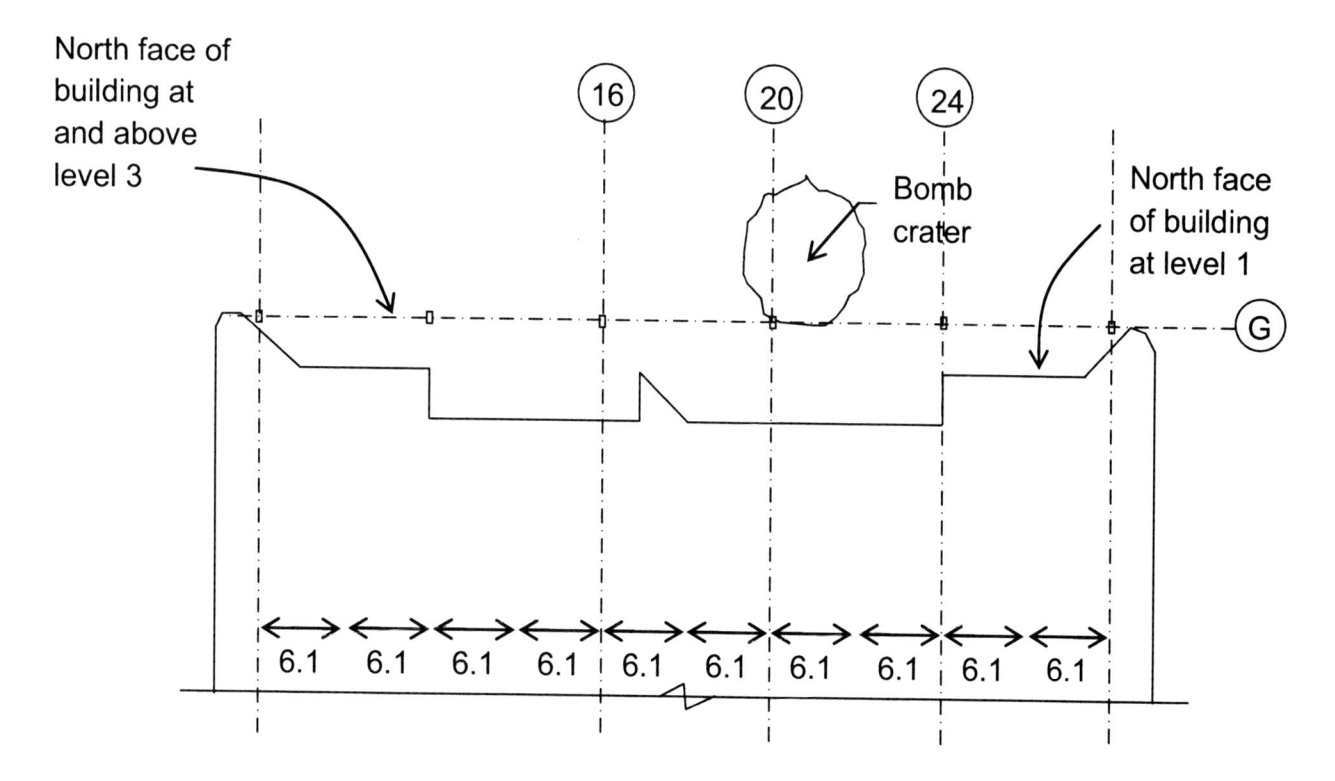

Figure 8.6: Plan of North of building at level 1 (Ground floor level)

Investigation

The blast was recorded on audio tape from a building nearby. The initial explosion was recorded as a noise lasting about 1/2 a second while the progressive collapse was a deafening roar lasting about three seconds. It was estimated that 90% of casualties were the result of crushing by falling debris rather than as a direct result of the explosion.

It was found that the building was constructed according to the drawings which were well above average quality. The cores and reinforcement samples indicated the concrete and reinforcement was as specified. The code specified lateral load due to wind was about 1.25 kN/m^2. There were no earthquake-resistant or blast-resistant requirements for the design.

Along the north perimeter the columns in upper levels were spaced at 6.1 m centres while below a large RC transfer beam at level 3 they were spaced at 12.2 m (Figure 8.7). There were no visible signs that any of columns had hinges at the top and bottom or that shear walls had hinged, thus collapse was not as a result of the building developing a failure mechanism from excess lateral load.

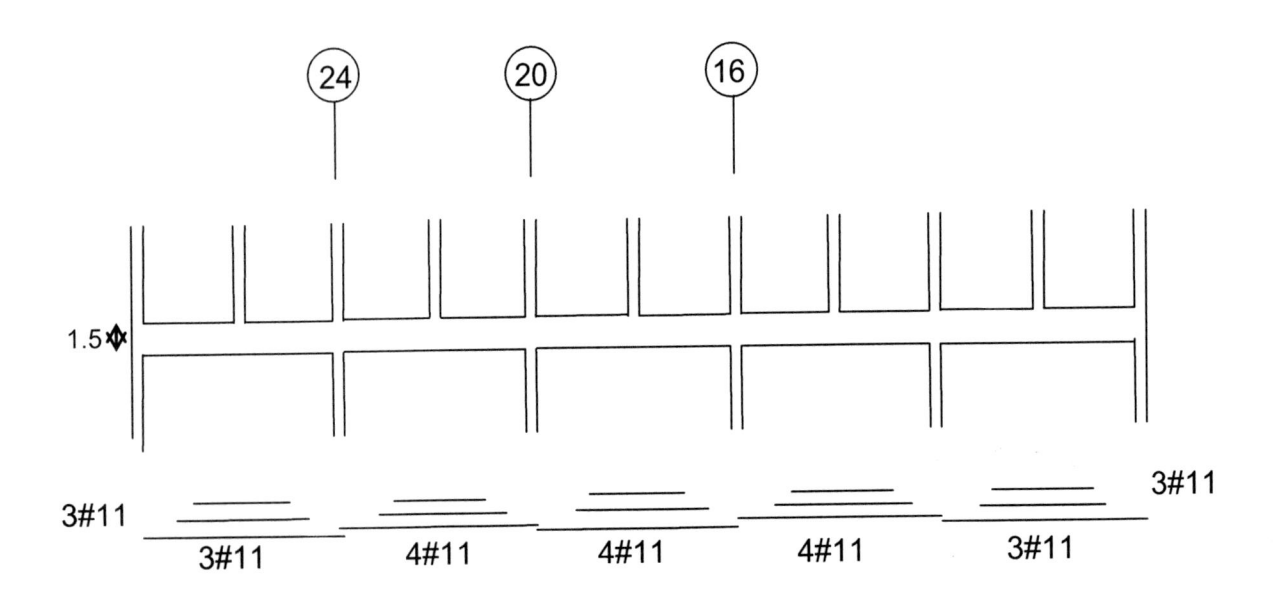

Figure 8.7: Transfer Beam at Level 3 on Grid G (bottom reinforcement shown)

Notice that the bottom reinforcement in the transfer beam was *not* lapped at the support, eliminating the possibility of catenary action. The transfer beam was found intact in the debris.

The detonation was only 4.75 m from column G20. G20 was 914 mm deep x 500 mm wide. It contained heavy longitudinal reinforcement and light rectangular links. It was reasoned that this column would have disintegrated as soon as it was hit by blast wave (the estimated pressure there was 69 MPa). Two other similarly reinforced columns (G24 and G16) would have been damaged as a direct result of the blast.

Some of the one-way slabs would have failed by hogging upward in response to the blast and then crashing back down 1 millisecond later as the pressure abated. However, the vast bulk of the destruction was due to the progressive collapse that followed blast.

It was found that even a "static" removal of column G20 at the ground floor was enough to cause the structure to become unstable simply because the transfer beam could not span 24.4 m as a catenary. It was estimated that minor detailing changes (such as lapping reinforcement in beams and additional shear links in columns) could have reduced the destruction by 80%.

Lessons to ensure blast resistance of special structures

(i.e., those where the risk of attack is considered to be high)

- Ensure there is proper continuity in transfer beams: allow for one support column to be removed without significant collapse of the structure.
- Protect the perimeter of the structure (no parking, loading bays, etc).
- Floors should be two-way systems.
- Use toughened glass on perimeter.

- Closely spaced link or spiral reinforcement in columns improves their blast resistance markedly.

Lessons for conventional structures

- Transfer beams can be a source of weakness. Ensure there is proper continuity in transfer beams. Use a higher Factor of Safety.

Aftermath

The event changed way the US federal government designs new buildings. A new federal office building was completed in 2004 on a site nearby. Blast resistant design was used for all elements. For the structure an "alternative load paths" approach was used: removal of any single column was explicitly considered and designed for. It uses a flat slab throughout. Bollards ensure that no vehicle is allowed within 15 m of the structure. Thick concrete walls protect perimeter. The cladding was designed to be blast proof too. Careful architectural design ensured the resulting building did not look like a military bunker!

8.5 World Trade Center 1&2, New York, USA, 2001.

[Chiles (2002), Gillespie (2002), Jennings (2004), Delatte (2008)]

WTC 1 and WTC 2 were known as the "Twin Towers". They were part of a seven building complex. The towers were designed by architect Minoru Yamasaki and engineer Leslie Robertson and opened in 1973. They were almost identical 110 storey (415 m) office structures. They were tube structures: wind forces were carried mostly by perimeter columns spaced at about 1 m centres, acting as a giant tube 63 m square on plan. The cores took most (60%) of the gravity load. They were briefly the tallest buildings in the world, until the opening of the Sears Tower in 1974. The structural plan is shown in Figure 8.8.

WTC 1 was known as the North tower and WTC 2, the South tower. The structural frame was steel. Instead the lift shafts and stairs were enclosed using four layers of 16 mm thick gypsum board known as "sheetrock". The floors were made of 100 mm light weight aggregate concrete (LWAC) on metal decking. Each tower had three stairwells, all situated in the core.

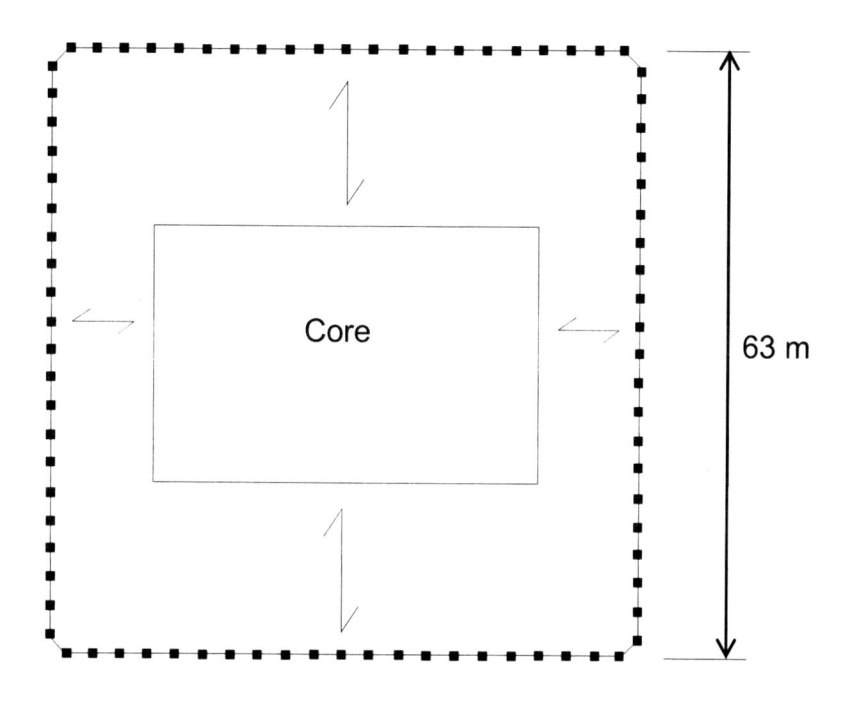

Figure 8.8: Plan of World Trade Centre (WTC 1 & 2)

(Each face of the building had 59 columns)

Large simply supported steel trusses 800 mm deep spanned the 20 m from the core to the perimeter columns. The floor deck was supported on transverse trusses (Figure 8.9). The steel was protected using a sprayed mineral fibre, to a two-hour standard to the open-web truss system and three-hour standard to columns.

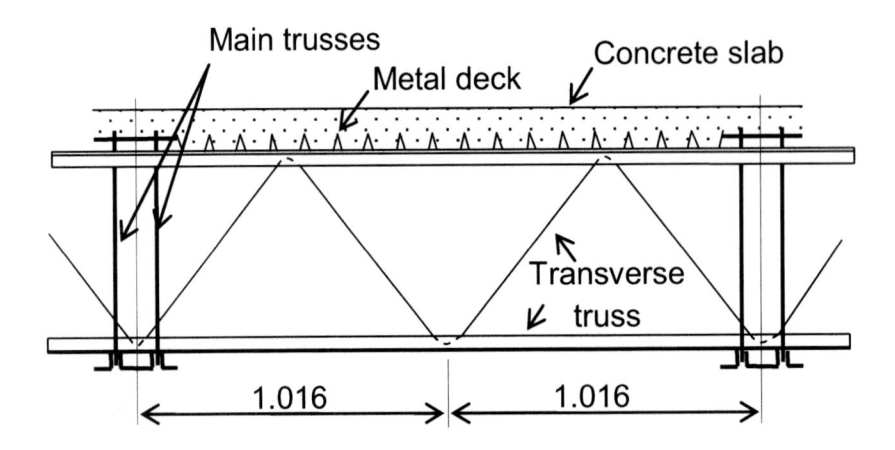

Figure 8.9: Transverse Truss (4 per main truss)

(Note detailing ensures transverse trusses could support the main trusses if the latter failed. The main trusses are in pairs improving robustness as the collapse of both is less likely.)

The towers were the first high-rise buildings to use a boundary layer wind tunnel in their design. Criteria had to be developed to assess the effect of wind-induced accelerations. Unlike most previously constructed tall buildings there was no masonry used in the construction, so there was no reserve to reduce accelerations and increase damping. Dampers were added to reduce the floor vibrations.

First Attack

At 12:18 pm on 26 February 1993 a large bomb exploded in the basement car park. The truck containing the bomb was on level B2 adjacent to WTC 1. The bomb was about same size as the Oklahoma bomb. The explosion caused the collapse of some floors in the basement but the towers themselves were not damaged (luckily no parking was allowed directly beneath towers). The public-address and emergency lighting systems failed. The towers filled with smoke. As a safety precaution both towers were evacuated using the stairwells. The evacuation took **four hours**.

The owners of the towers later spent over US$100 million to make physical, structural and technological improvements to improve fire safety. The improvements included installing fluorescent signs and markings and upgrading the emergency power facilities.

Second Attack

On 11 September 2001 both towers were attacked by aircraft. At 8:46 am WTC 1 was hit by a Boeing 767 flying at about 470 mph, centrally on the North face between floors 94 and 98. The Boeing 767 has wing span of 47.6 m, a length of 48.5 m and a maximum take-off weight of 180 tonnes. At 9:02 am WTC 2 was hit by a Boeing 767 flying at about 586 mph, on the South face near Southeast corner between floors 78 and 84.

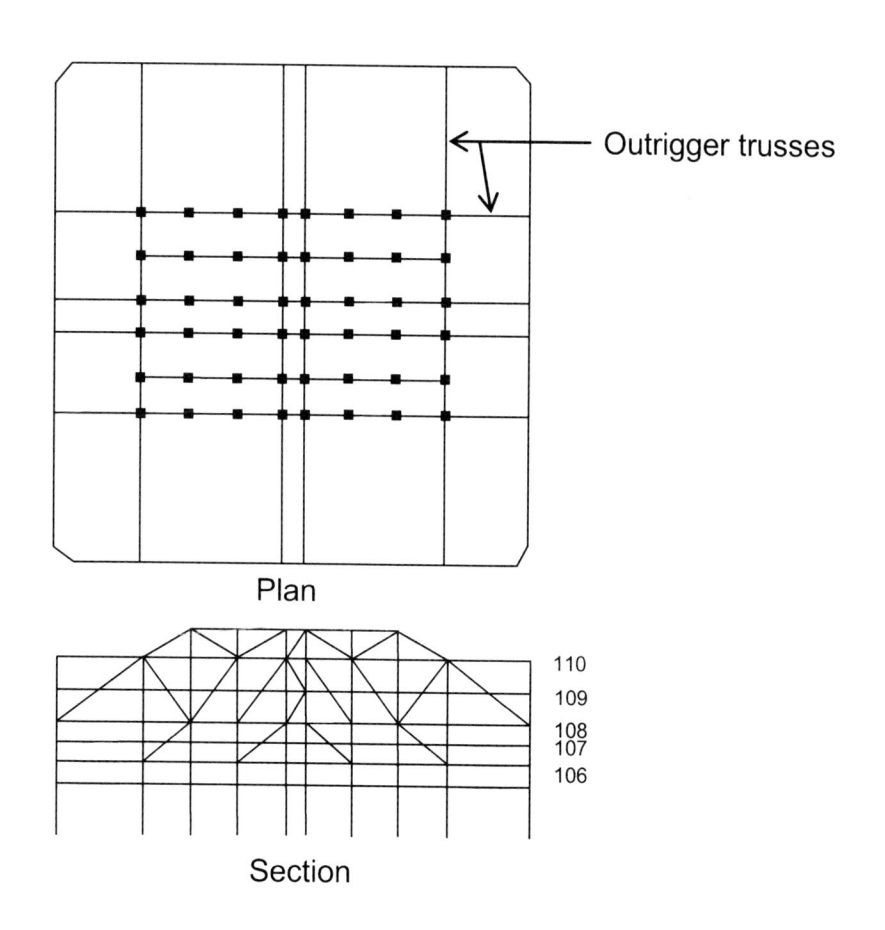

Figure 8.10: Outrigger truss system

It was later estimated that each tower rocked back 600 mm on impact. The aircraft sliced through 33 of the 59 steel perimeter columns of WTC 1 and 32 of them in WTC 2.

There was no immediate collapse: the surviving columns either side of the holes made by the aircraft with the assistance of outrigger ("top-hat") trusses in the roofs allowed an arch to form over the holes. The trusses are shown in Figure 8.10.

Each building had more than 1,000 times the mass of the aircraft, and had been designed to resist a wind load 30 times the weight of the aircraft as well as an unfueled Boeing 707.

Collapse

However, each aircraft contained over 38,000 litres of aviation fuel. Both were hijacked near the beginning of long-distance flights so were heavily loaded with fuel. Most of the aviation fuel is thought to have vapourised in the initial explosion. The sprinkler system was made ineffective by the impact and debris. Fires from office contents and the remainder of the aviation fuel burnt unchecked. They weakened the steel and reduced its modulus which led to buckling of columns. It is thought that the connections between the floors and columns failed too. WTC 2 survived for **56** minutes; WTC 1 survived for **103** minutes. 2,749 people died in twin towers. The dead included 343 firefighters who had rushed into the buildings soon before their collapse.

Investigation

The thickness of the sprayed mineral fibre fireproofing was in the process of being upgraded at the time of the attack. WTC 1 had a thickness of 38 mm on all members in the impact zone, while that of WTC 2 was only 19 mm. That and the fact that WTC2 was struck lower are thought to account for the different survival times.

It is thought that much of the fireproofing was dislodged by the impact. Had the fireproofing not been dislodged the temperature rise of the structural components would have probably been insufficient to trigger collapse.

Only 4 people from floors above the impact levels in either tower survived, as debris blocked many of the escape routes. The towers stood long enough for almost everyone (99%) who had an unimpeded route (i.e., below the impact zones) to get out. About 15,000 people escaped from towers WTC 1 and WTC 2 before they collapsed.

Many reported that the escape staircases were not wide enough to allow people to escape downwards and firefighters to climb upwards at same time; this delayed evacuation. There were three stairs, two 1,120 mm wide and one 1,420 mm wide. Survivors reported that shortly after the aircraft hit WTC 1 people in WTC 2 were told on building's PA system to remain at their desks.

Clearly, after the 1993 bombing the time required to evacuate the towers using the stairwells had improved substantially.

Lessons

- The tube structure can be astonishingly robust.
- At least one prominent engineer, Charles Thornton, advocates use of more concrete in new high-rise buildings especially to enclose lifts and stairs in a rigid box. Many new skyscrapers have this feature.
- Provide robust and durable **passive** fire protection.
- Treat active fire protection, e.g., sprinklers, as an addition to, not a substitute for, passive fire protection and do not consider it in extreme events.
- Provide robust adequately sized escape routes and diverse locations for them.

- Pressurize escape routes.

Many of these recommendations have been incorporated into Tomasetti & Abruzzo (2004) and The Institution of Structural Engineers (2002).

Additional Information

The risk of impact from aircraft is well known but rarely designed for. The twin towers were on the flight path to a nearby airport, and so were designed to resist impact of a Boeing 707 (with no fuel).

Most towers would not have lasted as long as those of WTC if subjected to a similar attack. It is likely that many more people would have died had the towers toppled over as soon as they were hit.

Many of most recent very tall buildings are constructed using concrete walls surrounding the core, with 8 outriggers stretching from the core to 8 large perimeter columns (which are often composite) to mobilize the resistance of the perimeter structure.

The maximum take-off weight of a Boeing 767 is 180 tonnes; for an Airbus A380 it is 583 tonnes.

"99% of all buildings would collapse immediately when hit by a 767"

Jon Magnusson (CEO of Skilling, Ward, Magnusson)

"The design of skyscrapers 50 storeys or more has been generally very robust, for 40 or less, it's not"

Dr Charles Thornton

"I don't think we should try and design tall buildings to resist the effects of aircraft crashing into them. I don't think we can solve the problem that way. The problem is with us, not our buildings, and it will be with us for a very long time."

Leslie Robertson (2002)

8.6 The Pentagon, Arlington, USA, 2001.

[Mlakar (2003), Delatte (2008)]

The Pentagon is the headquarters of the US Department of Defense. Construction was completed in 1943. It is one of largest office buildings in world (0.6 million m² of space).

The building is a five-storey *in situ* RC beam and slab structure. The concrete used had a cylinder strength of about 18 N/mm² and the steel a yield of 280 N/mm². The floors were designed for a live load of 7.5 kN/m². Floor slabs were 140 mm thick spanning 3 m onto secondary beams 510 mm deep. Primary beams were 660 mm deep typically with spans of 6.1 m. The ground floor columns were typically 535x535 mm and spirally reinforced. The exterior RC frames were in-filled with brickwork and faced with limestone. The storey height from ground floor to first was 4.3 m. A typical floor plan is shown in Figure 8.11.

Collapse

At 9:38 am on September 11, 2001 a hijacked Boeing 757 (weighing about 82 tonnes and containing about 20,000 litres of fuel) was intentionally crashed into the Pentagon. The aircraft hit the Southwest end of the building at the ground floor level at about 530 mph. 189 people were killed (including 125 people in Pentagon).

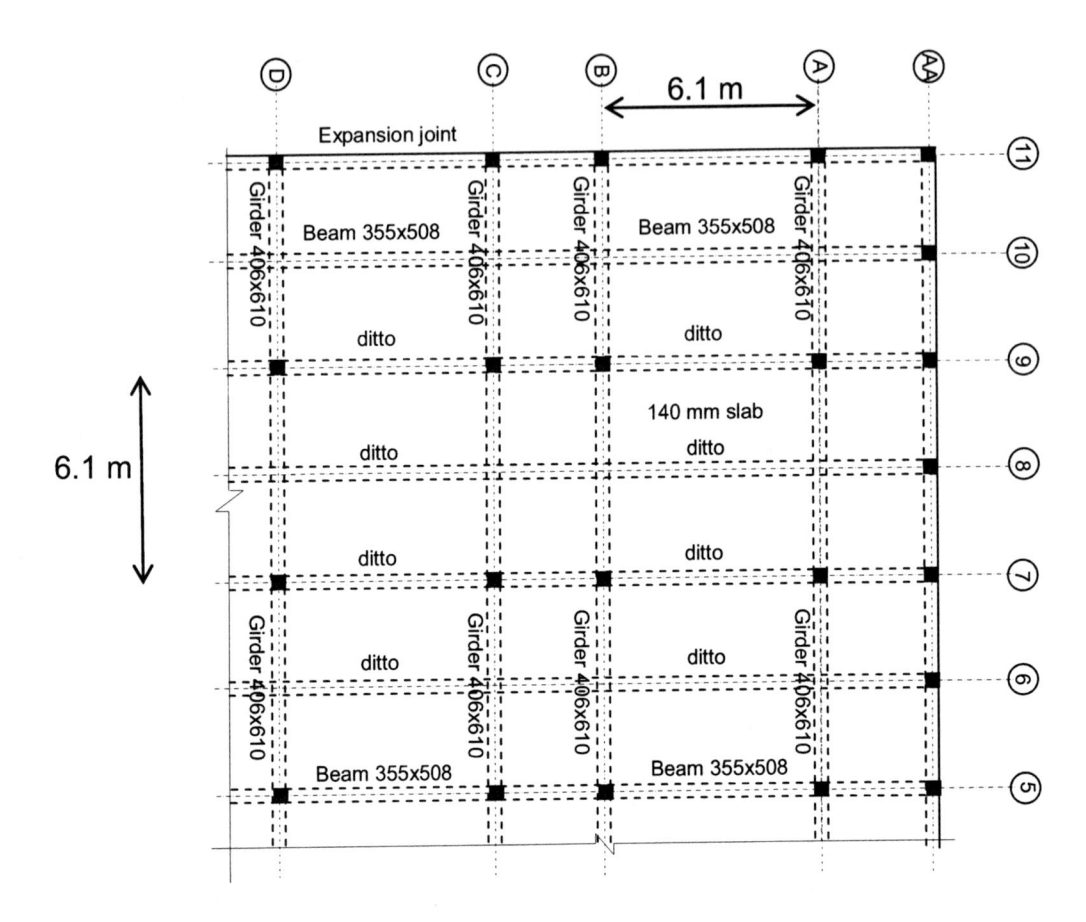

Figure 8.11: Typical floor plan

Investigation

The typical primary and secondary beam and column detailing was as shown in Figure 8.12, 8.13 and 8.14 respectively.

Link reinforcement in the beams was typically open-topped. Approximately half the bottom bars in the beams were made continuous by laps of 30 to 40 diameters at the supports. The design was economical for the time but would be regarded as conservative now. The columns were typically square with 6 longitudinal bars arranged in a circle enclosed by spiral links. (The code used in the original design allowed an increase in axial capacity for this reinforcement).

According to eyewitnesses the aircraft was flying only a few feet from the ground and apparently slid between the ground floor slab and the second-floor slab. Much of the wings (and thus fuel tanks) were severed as the aircraft penetrated the façade. The fireball was confined to the façade area. The aircraft frame was probably destroyed before travelling more than about 50 m into building.

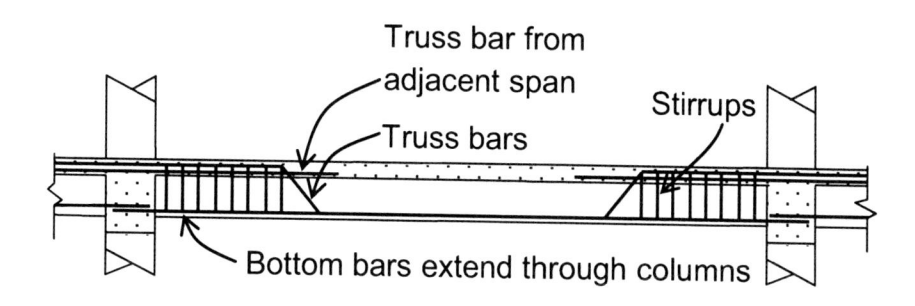

Figure 8.12: Secondary beam detailing

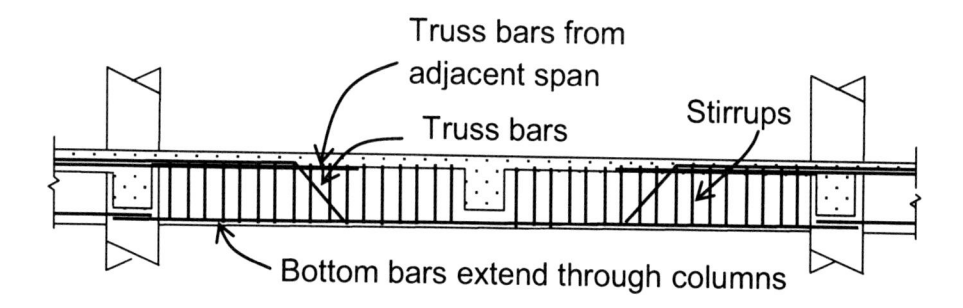

Figure 8.13: Primary beam (Girder) detailing

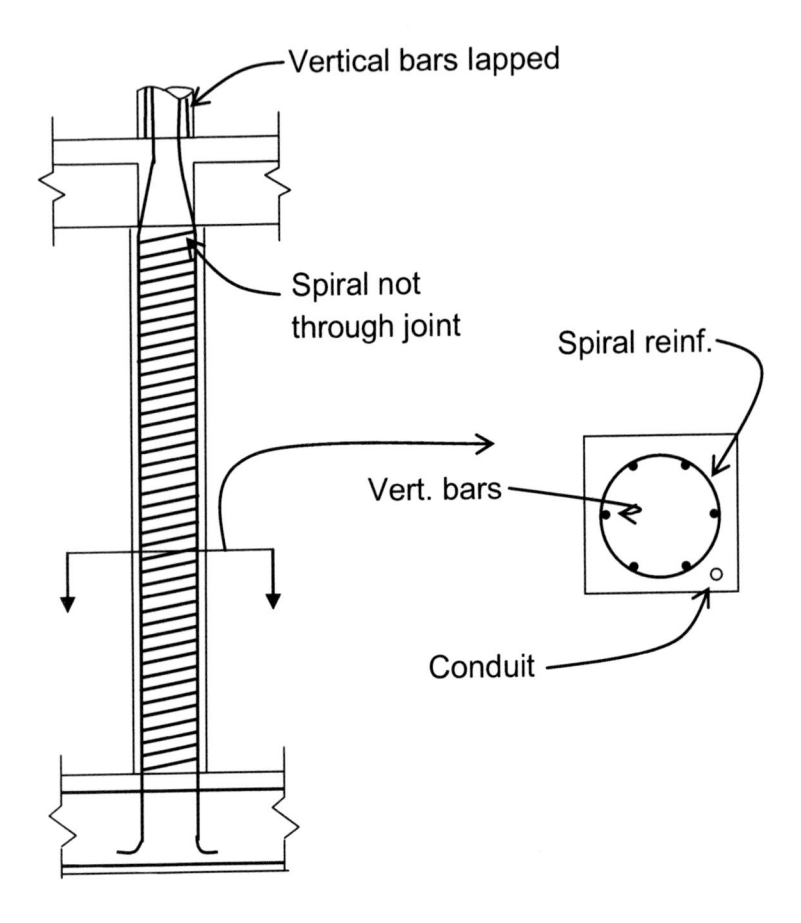

Figure 8.14: Column detailing

The fire damage was similar to that after a serious office fire. It was estimated that the fire had reached temperature of 950°C. It was found that 50 first storey and 6 second storey columns along the perimeter were lost in the impact (many bent not broken). Most of the serious structural damage was within a swathe of approx 25 m wide and extended about 70 m into the first floor of building.

The fire damage to the beams was limited to cracking and spalling. The spiral reinforcement on the columns was sufficient to ensure they would fail in **flexure** not shear so their energy-absorbing ability was very good. If the columns had been conventional tied columns it is likely that the collapse of the structure would have been much more extensive (Figure 8.15).

Structural Engineering Failures: lessons for design

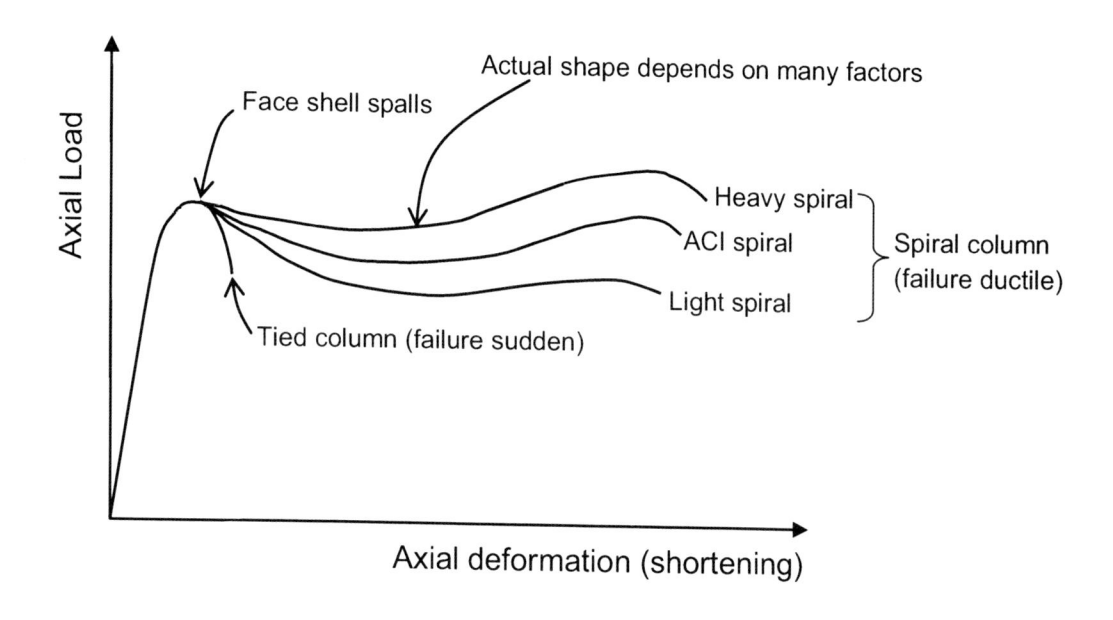

Figure 8.15: Typical column failure curves

Lessons

- The resilient structural system of the Pentagon meant that the damage was minimized. Attributes included the following:

    - Short spans between columns (to limit the remaining span if a column is removed)

    - Continuity (continuation of reinforcement, especially that at the bottom).

    - Redundancy (two-way floor system) and energy-absorbing capacity (spiral-reinforced columns).

    - Reserve strength (design LL was 7.5 kN/m^2 so floor and columns had spare capacity).

These lessons are also noted in Tomasetti & Abruzzo (2004).

"There is no question that the decisions made by the designers and the builders were critical in limiting the damage. Tied columns, optimized reinforcement quantities, inadequate splices, and longer spans would have led to total collapse of a larger area."

P. Mlakar (author of investigation report).

8.7   World Trade Centre 7, New York, USA, 2001.
[NIST (2008)]

The building was a 47 storey (160 m) steel-framed office building. It was constructed in 1985. Lateral loads were resisted primarily by the perimeter framing. The floors were concrete on trough decking working compositely with steel I-beams. Fire sprinklers were fitted. Steelwork had spray-on mineral fibre fire protection 12 mm thick, giving at least a 2 hr fire rating. The building design satisfied New York's Building Code.

Collapse

Impact and debris from the collapse of WTC 1 (10:28 am) started fires on levels 7-13 of the building and caused damage to seven columns on the south face. The building was evacuated; no attempt was made to fight the fires. The result was complete collapse of the building by about 5:20 pm. The fires had burned for seven hours. There were no injuries. It was the first ever complete collapse of a building primarily due to fires.

Investigation

The fires were mainly on the North and East side of the building. The sprinkler system failed to activate. Water was supplied from the water-main in the street outside and the water tanks on level 46. Floors 1-20 depended solely on the water-main supply for the sprinklers to operate. Above these levels water tank supply was available. The water main was severed during collapse of WTC 1 and 2. Thus fires burned unchecked on floors 1-20. The plan of a typical floor is shown in Figure 8.16.

The collapse was triggered by thermal expansion of the long-span floor beams on the east side of building. Their supporting girder (AB) framed into column 79. This connection was held horizontally by bolts. However, the beam slid off its seating once the bolts failed. The collapse probably began on level 13, initiating a "cascade" of collapses to level 5. The loss of restraint led to buckling of column 79 and that led to the collapse. Progressive collapse of the entire building followed.

It was estimated that thermal expansion due to temperatures of 400°C would have been sufficient to cause the failure of the connection.

It is likely that even without the impact damage to the south face, the fires alone would have caused collapse (as column 79 was not near the south face).

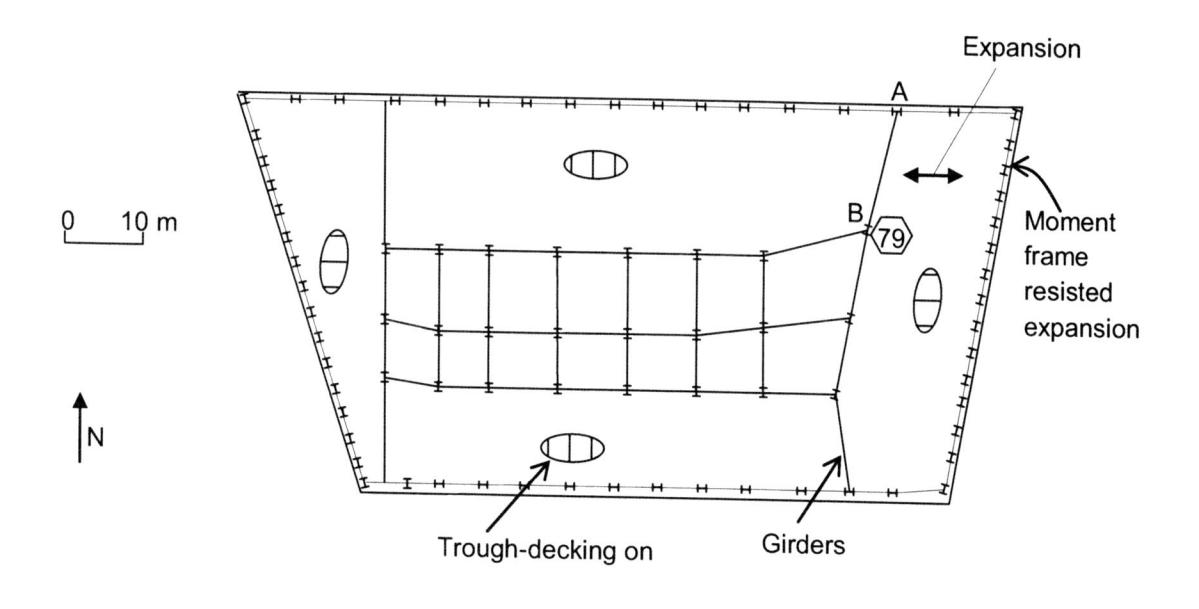

Figure 8.16: Plan of typical floor

Lessons

- The fire resistance of all buildings, particularly tall ones, should be carefully assessed. In particular, the response of the structural system to thermally induced movements needs to be assessed.

- Do not rely on active fire protection measures (sprinklers). Instead the structure itself should have sufficient passive resistance.

- Design so that failure of any one column does not lead to progressive collapse.

- Design connections to resist the forces induced by thermal effects.

- One-sided framing such as this means there are likely to be unbalanced expansion forces. This was made worse here by a lack of provision of shear studs.

This case study suggests that some building plans which cause less restraint are likely to be better in a fire. Two plans are shown below in Figure 8.17. The plan on the left causes less restraint to the floor slab (it expands away from the rigid core). The one on the right is likely to mean the compression stresses are locked in the slab. Spalling of the slab or even buckling might result.

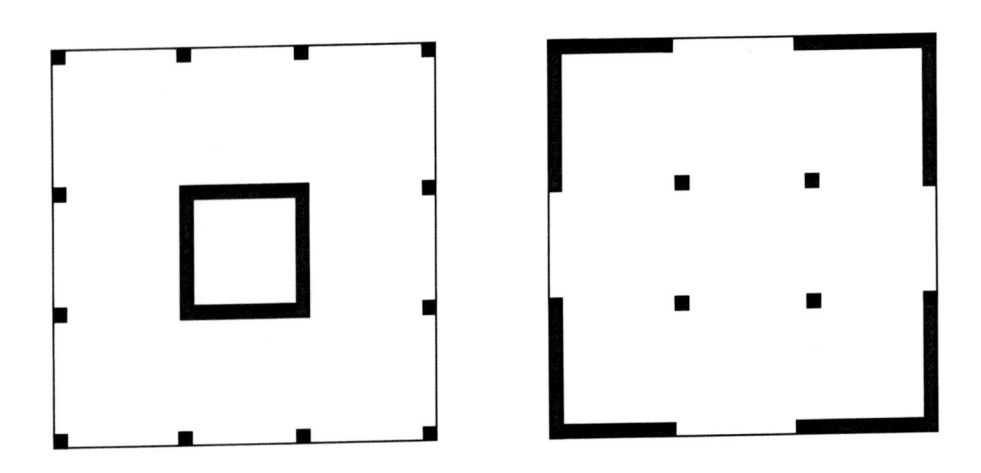

Figure 8.17: Floor Plans

Additional Information

In another case (Glanz 2001), fires started by the collapse of WTC 2 burned for nearly two days in a neighbouring 23 storey steel framed building. The building was constructed in 1907 (designed by Cass Gilbert, who also designed the Woolworth Building).

However, the columns were surrounded by 100 mm of ceramic tile as fireproofing. The typical floor span was 6 m. Except for a few spots where structural columns had slightly buckled the building was essentially undamaged. It is now refurbished and occupied.

Movement in a fire due to thermal effects can be 1-1.5% so an 8,000 mm span may move 80-120 mm (Ratay 2008).

Most modern steel-framed buildings are fire protected using a mix of mineral fibres and concrete-like materials as binders, and sprayed on. Even after it has been in place for years, it can be removed with a sharp tool without difficulty.

Chapter 9: Cladding

"We learn much more from failures than successes"

Henry Petroski, 1992.

- Case Studies
    - Stone: Amoco Tower, Chicago, 1989.
    - Render/tiles: High Rises, Hong Kong, 1992.
    - Brickwork: High rise, Singapore, 1993.
    - Concrete: "Big-dig" tunnel, Boston, 2006.

9.1 Amoco Tower, Chicago, USA, 1989.

[Kaminetzky (1991), Levy & Salvadori (1992), Feld & Carper (1997)]

Amoco Tower is an 82 storey steel framed tube building 342 m tall. It was constructed between 1971 and 1974. Its plan is a square of side 57 m. When constructed it was entirely clad with 1.27x1.14 m panels of white Italian Carrara marble 32-38 mm thick. (The technology for cutting such thin marble panels was then newly developed). The 43,000 panels were fixed to the structural frame using steel bolts.

Shortly after construction was completed many marble panels started to bow outward. For safety, these panels received additional bolts through to the frame. By 1988, 30% of the panels had bowed more than 13 mm and some as much as 38 mm. Tests on the panels showed that the marble had lost a large proportion of its initial strength.

Investigation

The method of fixing each panel involved bolts in <u>tight</u> holes thus resisting thermal movement of the panel. The temperature range in Chicago is about -35°C to +39°C. Thermal expansion of the panel would mean compressive stresses would be induced. The slenderness ratio (L/r) of panels was about 104 to 138, where r is the radius of gyration. (Most codes restrict this ratio to 120).

This is the correct method of fixing a panel so that temperature movements are not restrained (Figure 9.1).

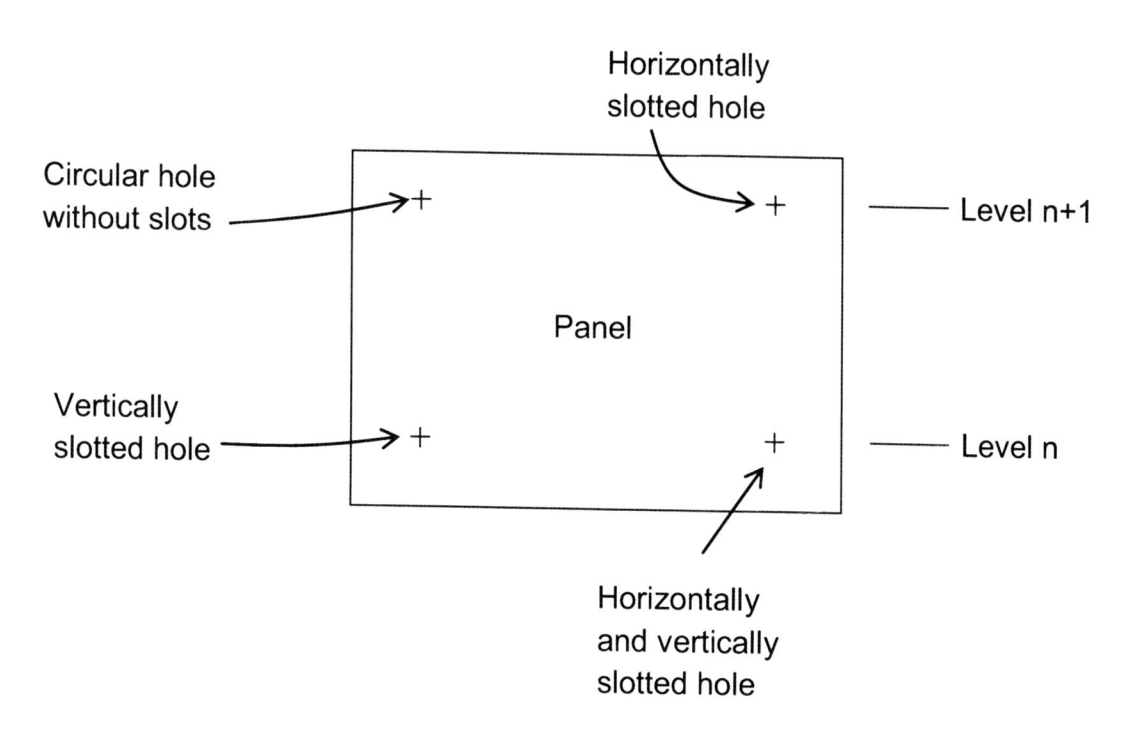

Figure 9.1: Allowing movement of panels

Lessons

- Many types of marble are prone to attack from temperature cycles, humidity and acid fumes of a polluted atmosphere (as is limestone, from which marble originates). The phenomenon is called "hysteresis".

- If affected, it can lose strength and can even buckle under its own weight.

- Many types of marble are thus not suitable for **external** applications.

- Slender panels, particularly in locations of large temperature variations, should be allowed to move without inducing stresses.

Aftermath

In 1989 the owners decided to substitute *all* of the marble panels with 50 mm thick granite panels. Besides being more resistant to buckling, granite is a granular rock mostly composed of quartz crystals and so "corrosion" is much lower. The cost of the new granite panels (US$75 Million) is estimated at more than half of the original cost of the entire building.

Additional Information

In Rochester, New York, a 42-storey tower was built in 1973. It was clad in 25 mm thick marble panels. In 1988, after numerous problems it was totally re-clad (in aluminium) at an estimated cost of US$35M.

In Stamford, Connecticut, a building was constructed in 1985. It was clad in limestone panels. In 1996, it was totally re-clad after its limestone cladding absorbed so much water that collapse was inevitable. The cost was US$5M.

9.2 Hong Kong Housing Authority apartment blocks, Hong Kong, China, 1992.

Mosaic tiles form the cladding to many apartment buildings in Hong Kong. Typically they are placed in sheets on to the rendered in situ frame. Unfortunately in some cases the cladding de-bonded and fell, sometimes from the top of blocks that were up to 30-storeys high. In several cases storey high panels fell. Luckily nobody was injured.

Investigation

The RC frame had a thin ceramic tile finish with a sand/cement render backing. There was no obvious pattern in failure locations, e.g., no correlation of failed panel to sunshine or even time of construction. The construction materials used were generally good.

As is the case with many similar problems (Ransom 1981), it was considered that poor workmanship in installing the tiles was the likely cause of the problem here, with a design deficiency exacerbating the problem:

- Little effort had been made to ensure a good mechanical key between the concrete and the render, e.g., by coating the surface of the substrate with "spatter-dash".

- The dimensional control of the RC frame was poor, so the render thickness was often required to be over 50 mm, while its expected thickness was less than 20 mm.

- In some cases the formwork used in constructing the original RC had not been properly removed.

- It was likely that the mix actually used for the render was too rich (i.e., used too much cement and not enough aggregate). Thus there was excessive shrinkage of the render, eventually causing the bond with the substrate to fail. (It would have been better to use a mix of 1 part cement to 6 parts sand.)

- There were insufficient vertical and horizontal movement joints which allowed stresses to build-up in the render/tile finish.

Repair

Various alternative methods including thermography were evaluated for locating the de-bonded areas. Hammer-tapping or hammer dragging was found to be the most reliable. There was little alternative but to cut out and replace the de-bonded portion.

Lessons

- Workmanship in installing a tiled finish is crucial.
- Surface of the substrate must be properly prepared (i.e., carefully cleaned and proper "spatter-dash" applied or the surface scored while still setting).
- Movement joints to allow the render to "breath" are necessary.
- Tiled finishes are often avoided, especially in high-rise buildings, for these reasons.

9.3 Condominium Building, Singapore, 1993.

This case study concerns a 20-storey condominium building near the Eastern coast of Singapore. It was built in the 1980s. The RC frame was clad with clay brickwork.

Collapse

One day in 1993 without any warning, a storey-high panel of brickwork bulged and then popped out from a lower storey. Nobody was injured.

Investigation

It was clear from brief inspection that there had been no provision for movement of the cladding. The sources of movement are as follows:

- Gravity load on the concrete means that creep of the frame is downwards (creep is roughly proportional to stress).

- There also would have been some frame shrinkage, which would have resulted in downward movement of the frame too.

- It is likely that the installation of the brickwork began before the frame was completely constructed, so in addition to creep and shrinkage there was elastic compression of the frame to be allowed for.

- Brickwork responds to temperature changes. The surface can reach temperatures of 70°C in Singapore (Chew et al. 1998).

- Brickwork undergoes irreversible expansion as its moisture content increases once it is removed from the kiln.

There were no horizontal joints present to allow the brickwork to move vertically independently of the concrete frame. Thus the brickwork was forced to take axial load, resulting eventually in it buckling

The combination of a brickwork skin and a concrete frame has in the past often been found to be troublesome. In addition, in this case the brickwork support detail made the problem worse as any vertical load was forced to be eccentric to centre of wall. The brick slips were probably squeezed out first. See Figure 9.2.

Cladding

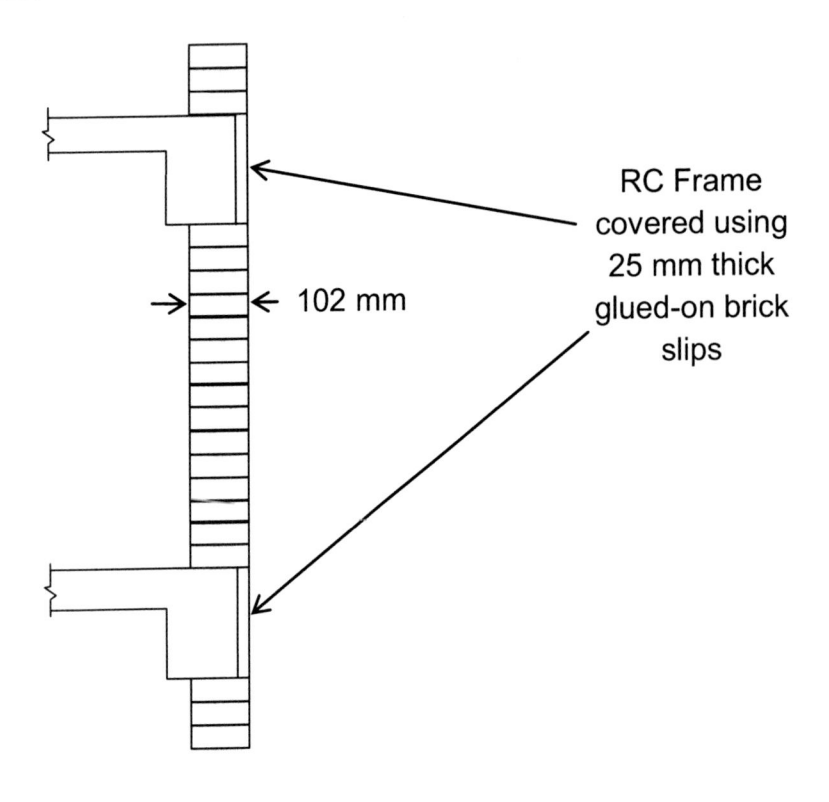

Figure 9.2: Cladding detail at a Singapore Condominium

RC Frame covered using 25 mm thick glued-on brick slips

102 mm

Repair

It was recommended that new horizontal joints be saw-cut into the brickwork at each storey. This must be done carefully as much energy may be released. Often the newly saw-cut joint closes as the brickwork releases its stress.

A stress-relief test can indicate *in situ* stresses that the brickwork is carrying. A stress-relief test may be carried out as follows:

Mechanical gauges are first placed in a vertical line above and below the brick that is to be removed. The relative positions of the gauges are noted. The brick is then removed. A flat jack is inserted. The jack is inflated until the gauges return to their original positions. The force required to return them indicates the existing level of axial stress in the brick.

Lessons

- Allow adequate horizontal movement joints to cater for differential movement that occurs between brickwork and concrete.

- Brickwork distress is usually worse where the brickwork clads an RC frame rather than a steel frame.

- Delay use of bricks to ensure their moisture expansion is complete if possible.

- Distress often appears soon after construction, especially if the construction is fast-track (i.e., the brickwork installation begins before the frame is completed).

Additional Information

[Ransom (1981), Ross (1984), Kaminetzky (1991), Feld & Carper (1997), Alexander (2001), Campbell (2001), Newman (2001)]

There have been numerous cases of omission of this detail throughout the world.

On some projects the structural engineer receives a fee for ensuring the cladding details are satisfactory. More thoroughly burnt bricks (i.e., darker) are less prone to irreversible expansion. Brickwork expansion is more pronounced in climates with persistent driving rain (e.g., Northern Europe and North America).

Details that do not cause problems on a relatively low-rise building may be troublesome on a high-rise building (since creep and some elastic deformation have to be accommodated). E.g., external brickwork panels built tight to the concrete frame usually cause no problems if building is relatively low (say up to about 10 storeys in climates such as Singapore's), but are likely to cause problems (e.g., vertical cracking/bulging in the brickwork) on taller buildings.

Shortening of the concrete frame due to this can be around 5 mm per storey. Thus the cladding must incorporate movement joints which allow for at least 5 mm of movement per storey, and about 6 mm in case of brick. So the joint filled with sealant is typically 20 mm wide.

9.3 "Big-dig" tunnel, Boston, USA, 2006.

[Angelo (2007), Miga (2007), Delatte (2008), Salmon (2008)]

The project began in 1991. It consists of a new underground road, 2 bridges, and an extension of the I-90 highway to the near-by airport requiring an immersed tube tunnel across Boston harbor. The whole project cost US$14.8 billion.

The cut-and-cover tunnel approaches to the immersed tube tunnel were completed at the end of 1999. The tunnel had a false ceiling made of RC panels (for aesthetics and to facilitate air-flow). Each RC panel, weighing up to 2-3 tonnes, was hung from steel supports which were in turn hung from the *in situ* concrete roof using cables and end-plates fixed using post-drilled epoxy anchors 200 mm long. See Figure 9.3.

Collapse

Suddenly in July 2006, 10 panels of concrete fell as well as their steel supports and 20 anchors (the total weighing about 26 tonnes). One person was killed and one injured.

Investigation

Clearly the fixings had pulled from the *in situ* ceiling. Thus the investigation focused on the fixings. The Contractor used a "fast-set" epoxy. This epoxy had acceptable short-term strength but it was incapable of supporting even much lower loads over an extended period of time, i.e., it had poor resistance to creep. This was a "creep-rupture" failure. Short-term pull-out tests were carried out on-site at the time of construction. The anchors passed the tests.

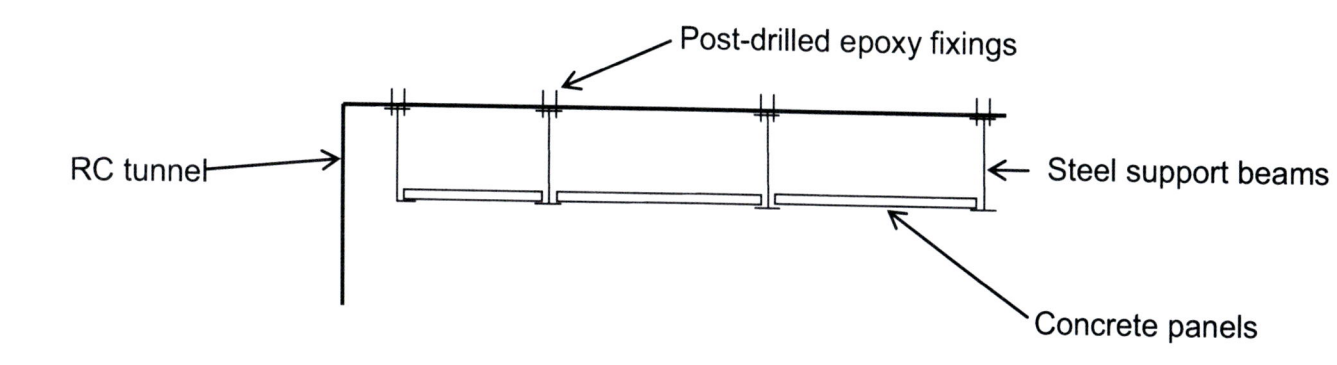

Figure 9.3: Fixing of suspended ceiling

The anchors were installed 6½ years before the failure. The epoxy supplier tested its "fast-set" and "standard-set" epoxy in 1995 and found "fast-set" epoxy had poor creep resistance (it failed within 80 days). However, the supplier's product information did not clearly say that the "fast-set" product was unsuitable for sustained loads.

Both types of anchor were supplied to the project. The supplier maintains that the Contractor was clear that only "standard-set" anchors were to be used for ceilings. However, the Contractor used "fast-set" for wall panels and by mistake for ceiling panels too.

Lessons

- Require regular mandatory tunnel inspection and development of protocols to test adhesive anchors used to hold tunnel ceiling panels. (Inspectors would have seen gradual pulling-out of anchors well before eventual failure.)

- Ensure the testing is representative of the eventual use of the product: a long-term test was required. Short-term tests can be misleading.

- It is my opinion that the use of an epoxy anchor for overhead application in a tunnel is inappropriate. Some epoxies lose their effectiveness at 60^0C and most by 150^0C. A tunnel fire can reach temperatures of well over 1000^0C so the application was risky.

Aftermath

It was found that 1,100 epoxy anchors used on the project were faulty. As a temporary measure, additional mechanical undercut anchors were installed. Research predicted even standard-set anchors would not meet the 100-year service life of the tunnel.

The National Transportation Safety Board (NTSB) later banned epoxy from overhead applications. The tunnel was reopened in June 2007.

Chapter 10: Miscellaneous Failures

"Hindsight is always twenty-twenty"

Billy Wilder (1906-2002)

- Case Studies
    - USAF warehouses, Ohio/Georgia, 1955/6
    - Stepney School, London, 1974.
    - Kemper Arena, Kansas, 1979.
    - Station Square Shopping Centre, Canada, 1988.
    - K-B Bridge, Palau, 1996.
    - Millennium Bridge, London, 2000.
    - I-35W Bridge, Minnesota, 2007.
    - Convention Centre, Pittsburgh, 2007.

10.1 US air force warehouses, Ohio/Georgia, USA, 1955/6

[Elstner & Hognestad (1957), McKaig (1962), Shepherd & Frost (1995), Feld & Carper (1997), Delatte (2008)]

Two almost identical warehouses of dimensions 610 x 122 m were constructed around 1954. They were *in situ* RC rigid frames consisting of 6 spans of around 20 m. The frames were at 10 m centre to centre. The beams were typically 915 mm deep with haunches at the supports making the depth 1370 mm. It was intended that there were expansion joints each 61 m or so. A typical elevation is shown in Figure 10.1.

Collapse-Ohio

Cracking was noticed two weeks before failure near the point of contraflexure. Shoring was installed. Cracks were diagonal starting at the top of the roof beam about 0.45 m past the end of the cut-off of flexural reinforcement. There was a sudden collapse of the roof on 17 August 1955. No other loads apart from self-weight were present.

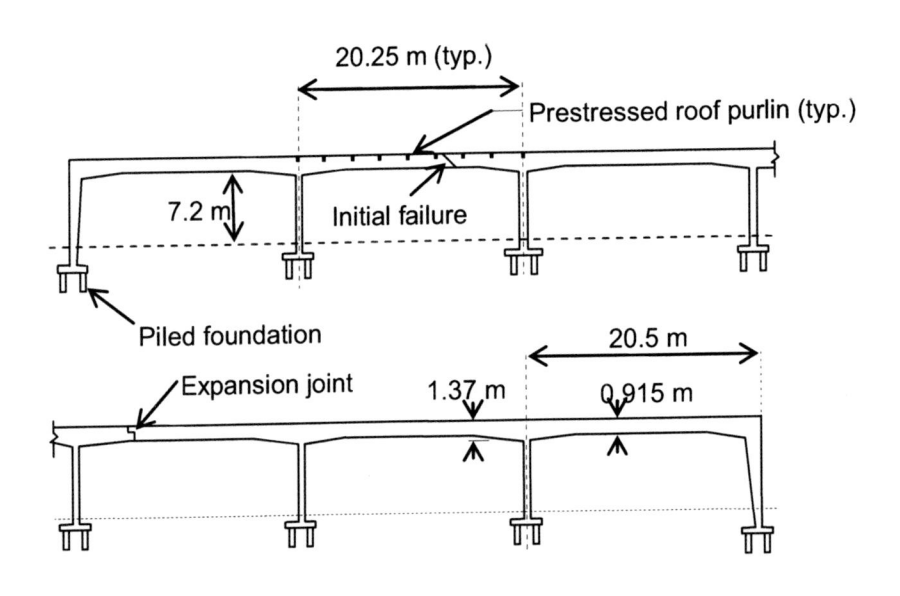

Figure 10.1: Typical Elevation of frames

Collapse-Georgia

Similar diagonal cracking was noticed shortly before failure. Crack widths were up to 13 mm. Shoring was installed. There was a complete collapse suddenly on 5 September 1956.

Investigation

The frame designs were similar. The RC frames were designed in 1952 to the then current code (ACI-318:51). Shear links were placed near the supports only. No shear links were placed beyond the point of contraflexure. Top flexural steel terminated completely near the point of contraflexure.

Each frame was constructed within a day (thus shrinkage was likely to have been high). In both cases expansion joints were locked (several other warehouses were built to similar designs, but incorporated (working) expansion joints at 61 m centres). Large temperature variations were experienced.

Laboratory tests reproduced the failure (Figure 10.2). Investigators concluded that "failure took place by a combination of diagonal tension (shear) due to dead load and axial tension due to shrinkage and temperature change".

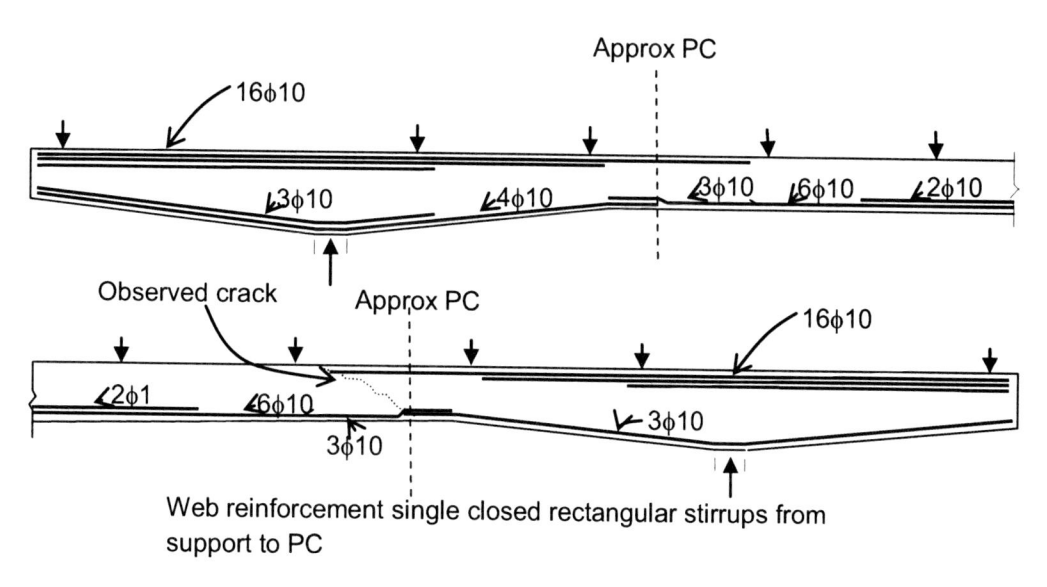

Figure 10.2: Details of laboratory test specimens used to investigate failure (reinf. to **ACI-318:51**)

It is interesting to note that the bottom reinforcement is crucial to preventing a brittle shear failure from being a catastrophic one too (i.e. it could prevent the roof from falling). The action is similar to that of the bottom bars in post-punching. However in this case the effectiveness of the lap in the bottom reinforcement is questionable. The compression zone (i.e. the lower part of the beam) will fail due to shear and be heavily cracked. Thus the concrete surrounding the lap will lose strength causing the lap to lose effectiveness. Thus after shear failure has occurred there is nothing to stop the beam from falling.

Research by Kotsovos & Pavlovic (1999) would suggest that the point of contraflexure is a vulnerable part of the beam. Link reinforcement is essential around this point.

Lessons

- These two cases illustrate the importance of providing at least some minimum amount of reinforcement, (known as temperature steel), to resist tension forces due to thermal, shrinkage, and other effects.

- Real structures may not behave in the same way as our simplified models, and develop forces and stresses where our analyses suggest there should be none.

- Top flexural steel should extend, preferably at least an anchorage length, past the point of contraflexure.

- Restrained axial shrinkage may cause a crack at or near the point of contraflexure so it is essential, especially in long structures, that links are provided here.

- Expansion joints (which consisted here merely of two steel plates) must be properly detailed to allow expansion/contraction.

- The point of contraflexure is likely to be a weak point. The point of contraflexure region is in direct tension and so is further weakened by axial compression.

- This feature of continuous beams is little known, as in research simply supported beams are most often tested.

- The detailing of the bottom reinforcement is important. To ensure resistance after shear failure has taken place no lap should be placed in these regions. Instead, the lap should be centred directly over the support.

Aftermath

Existing warehouse frames were strengthened by adding external shear reinforcement consisting of tensioned steel strapping and steel angles at the lower corners of the girders.

The ACI Building Code shear provisions were also revised (requiring increased reinforcement).

10.2 Stepney School, London, UK, 1974.

[Bate (1974)]

The building in question is a single storey swimming pool building 21 x 10 m. The short direction was spanned by 14 precast prestressed I-beams made using High Alumina Cement (an accelerator popular with precasters at the time). Bearing for the I-beams was 100 mm onto *in situ* RC edge beams. The building was constructed in 1965. Longitudinal and transverse sections are shown in Figure 10.3 and 10.4 respectively.

Collapse

The roof suddenly collapsed on 8 February 1974. The first signs were noticed at 10:45 am when students noticed spalled concrete in the pool water. Cracking then appeared in one roof beam near mid span. The hall was evacuated. A single I-beam collapsed shortly afterwards. Later that day another beam fell.

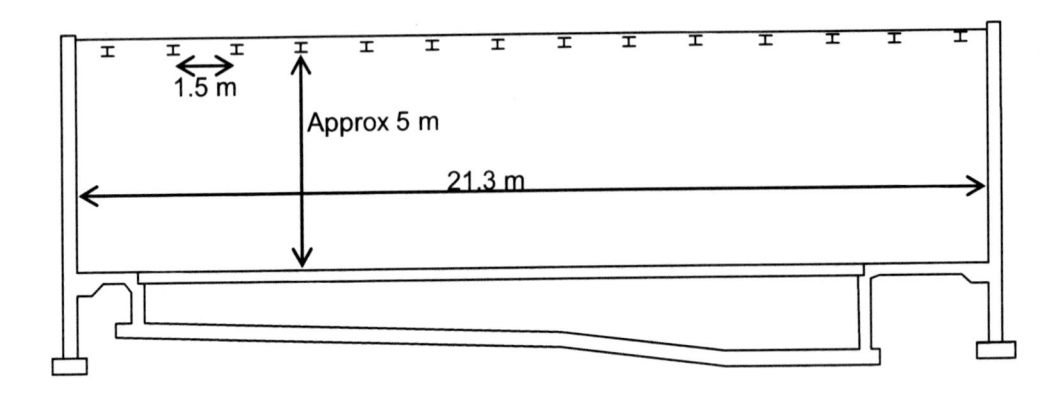

Figure 10.3: Long section through building

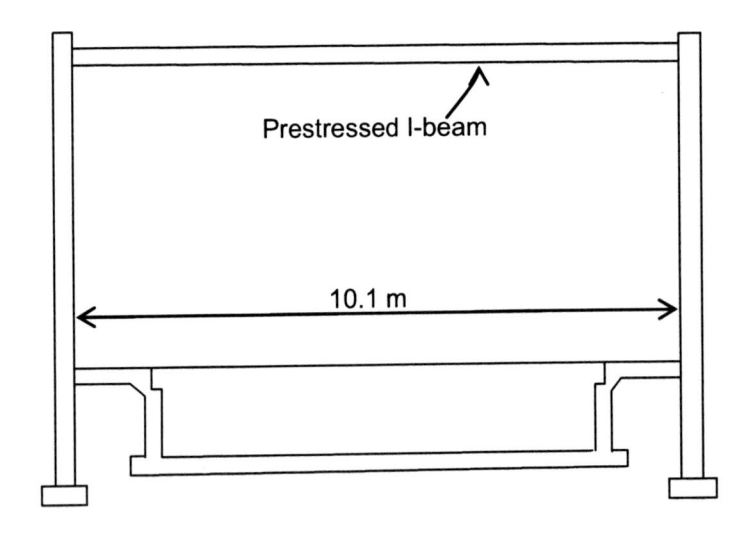

Figure 10.4: Cross section through building

Investigation

The previous June, the roof of the Assembly building of Camden school collapsed. There too the roof incorporated precast beams made using HAC. This resulted in inspection of this structure (September 1973). That inspection was visual with the aid of a rebound hammer. No evidence was found of deterioration of the concrete strength (the measured minimum equivalent cube strength was 52 N/mm^2).

The tendency of concrete made using HAC to "convert" (i.e., reduce in strength) was known. An Institution of Structural Engineers report of 1964 warned about it. However the precise conditions under which it was likely were not clear. It was known however that concrete might convert over time resulting in weakening of concrete when the service temperature was over 27°C.

It became clear later that moisture and high temperature worsen this tendency dramatically. In this swimming pool roof we had both; the uninsulated roof of the swimming pool meant an average temperature of over 27°C was likely with constant high humidity. The resistance to chemical attack was known to fall also.

The concrete in the collapsed beams was weak (typically 16 N/mm^2 measured from cores, having been 79 N/mm^2 at 28 days). The fractured surfaces of concrete also showed crystals of ettringite indicating concrete has been attacked by sulphates (the gypsum plaster probably the source of sulphate). There was no corrosion of the embedded steel.

Lessons

- Introduce innovations only after thorough testing. The use of HAC in the UK is one of several failures due to innovation incorrectly introduced before sufficient testing.

- Other examples are the use of calcium chloride as a concrete accelerator and the use of "high-bond" mortar additive both of which caused corrosion of embedded steel.

Aftermath

The use of HAC is now banned in most countries for structural applications. However, it continues to be used in non-structural refractory applications due to its high temperature resistance (it can resist temperatures up to 1350^0C).

10.3 R. Crosby Kemper Memorial Arena, Kansas, USA, 1979.

[Ross (1984), Levy & Salvadori (1992), Schlager (1994), Shepherd & Frost (1995), Feld & Carper (1997), Wearne (2000), Jennings (2004), Delatte (2008)]

The arena was constructed in 1973. It is a 130 x 94 m building with a seating capacity of 17,600. The design won an award from the AIA. The architect was Helmut Jahn (of Murphy/Jahn).

The roof consisted of 50 mm concrete on steel trough decking placed on "open web joists" at 2.7 m spacing. In turn the "open web joists" were supported on seven secondary cantilevered trusses (spanning N-S) at about 16.4 m spacing. Each of these secondary trusses was hung in 6 places, using "pipe hangers", to 3 large external triangular space portal frames 9 m deep and of span 108 m each (spanning E-W). This resulted in the interior of the Arena being column free. Thus the roof was suspended from 42 hangers. The pipe hangers were connected to the bottom chords of each of the triangular trusses using 4 bolts each. Figure 10.5 shows a plan.

Collapse

On the evening of June 4, 1979, about 108 mm of rain fell in 2 hours accompanied by a North wind gusting to 32 m/s (70 mph). This was about a one year storm for the area. Progressive collapse of central portion of roof took place without warning (a 60 x 65 m section). Luckily the arena was empty. Twenty-four hours before, thousands attended an AIA convention. Two days before that stadium held 13,200 people.

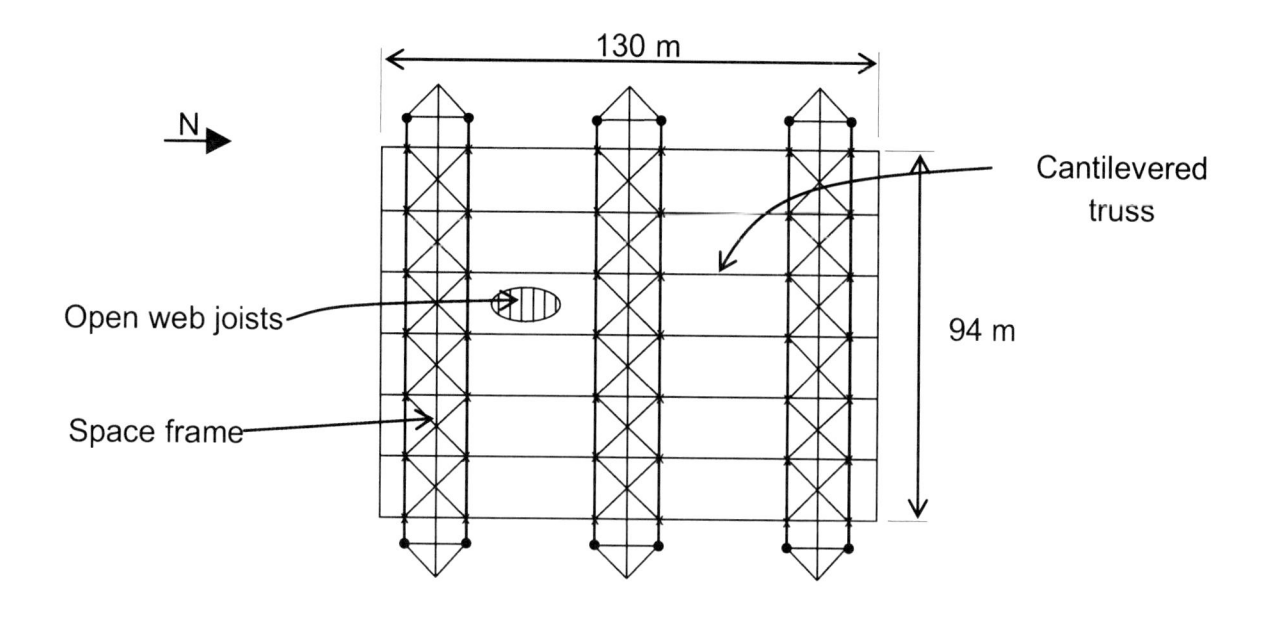

Figure 10.5: Plan of structure of Kemper Arena

x - hanger point

Investigation

The main space frames remained intact. Thus it was clear that the connection between roof and long-span trusses failed. The hanger connection is shown in Figure 10.6. To guarantee good contact between the hanger's welded end-plates and the truss and also to prevent a "cold-bridge" situation, the designer placed a 6 mm thick plate made from micarta (a plastic) between them. (The hanger's endplate was welded, so some distortion was expected).

Failure began at a single hanger and progressed to the others along the space frame. Four 35 mm dia. bolts connected each roof truss to a pipe hanger. It was estimated that during the six years preceding the failure the connection was subject to 24,000 oscillations by the wind.

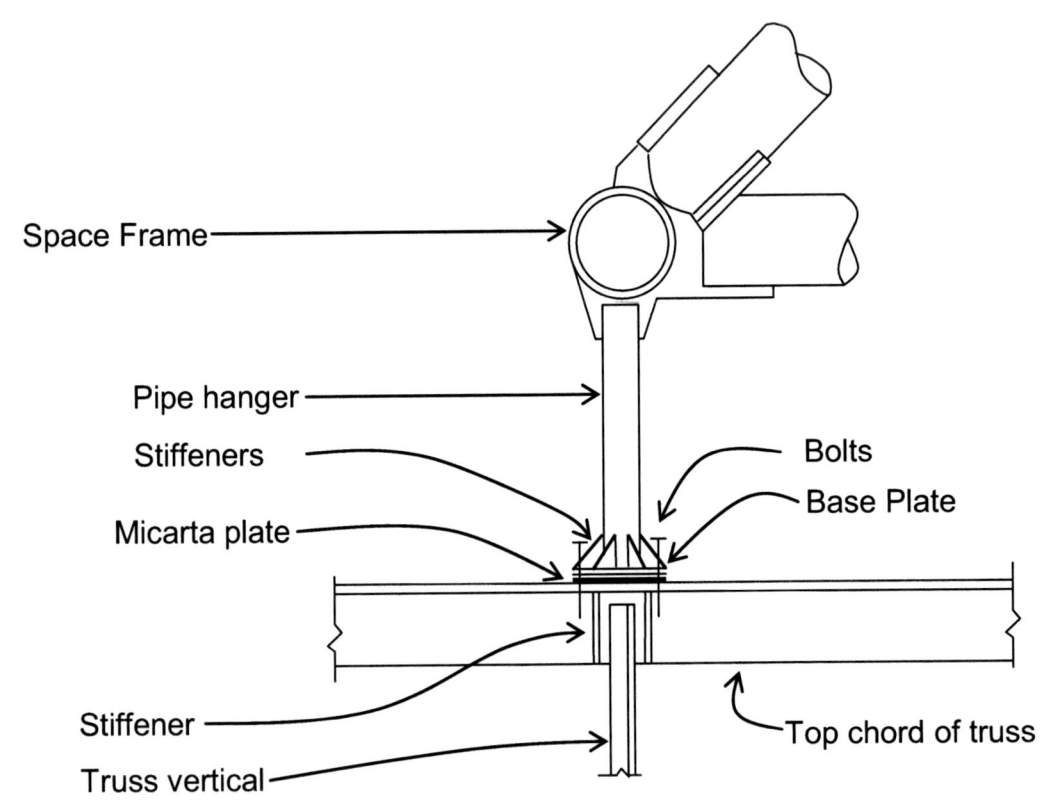

Figure 10.6: Detail of hanger connection

The bolts used were the high strength pre-loaded type (similar to HSFG bolts). Pre-loaded bolts are normally very good if the load is fatigue. However, the presence of plastic plate is very significant.

Just as in prestressed concrete the concrete creep results in a loss in extension of the tendon and so a loss of prestress, the plastic plate creeps under sustained load and so gradually the pre-load is lost.

The bolts were made from high-strength steel (yield stress f_y = 640 N/mm^2). This steel is relatively brittle. Fully untightening and tightening these bolts about 5 times reduces failure load by about 65%. Thus under dynamic conditions the capacity of the untorqued bolts would have dropped dramatically. Under design loads (considered static), the factor of safety was quite high.

The dead weight of the roof was 1.3 kN/m^2, and it was designed for a uniform live load of 1.25 kN/m^2 due to rain, etc.

Although each cantilevered roof truss was connected in 6 places by 4 bolts at each place, it was still statically determinate as the "suspended" span was pinned at each end. Clearly if a single hanger failed, e.g., because of bolt fatigue, this truss would become unstable (initial failure was at hanger 1 shown in Figure 10.7). In fact it could be shown by simple calculation that if a single hanger failed, the neighbouring hangers would fail too. Thus the roof design lacked redundancy.

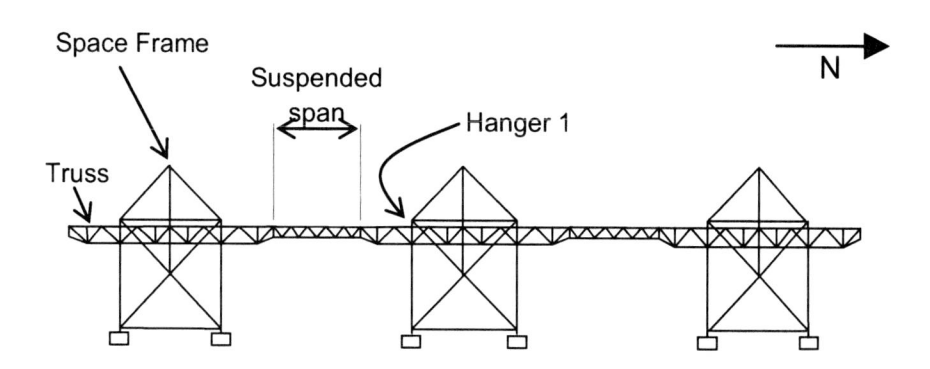

Figure 10.7: Configuration of main trusses at Kemper

Thus the initial investigation focused on the bolts themselves. However there was another factor (noticed later) that would account for the failure: the roof was <u>too flexible</u> to resist water ponding and so was unstable!

It was intended that the roof should not drain quickly as it might overload the local sewerage system. Thus there was a deliberate lack of drains on the roof. The 12,000 m^2 roof had only eight drains, i.e., one/1500 m^2. (One/186 m^2 was recommended by the local code to ensure rapid discharge). Scuppers along the perimeter of the roof were set 50 mm above the roof level. Water accumulation on roof was aggravated by wind: gusts from the North pushed water towards the South end.

Calculations showed that the roof was unstable. It did not have enough stiffness to prevent instability as a result of ponding. It was estimated that actual depth of water on the roof was up to 230 mm (i.e., 2.3 kN/m^2). Recall that it was designed for a LL of 1.25 kN/m^2. This would cause failure by overloading of bolts. Thus even if the capacity of the bolts had not been reduced by any wind-induced fatigue (and no doubt the capacity was), they were stressed much more highly than assumed in the design.

We are used to assuming the deflection has no influence on the load; but if the structure is a flexible roof, that is not true! Referring to Figure 10.8: Load => roof deflects forming dish-shape. If dish then fills with water there is more load and so more deflection, etc.

At Kemper the drains were located in the centre of the roof. The roof was sloped 100 mm towards these drains (thus the fall was about 1/450). The roof was designed to take only the ponding implied by Figure 10.9.

Elastic deflection
forms dish shape

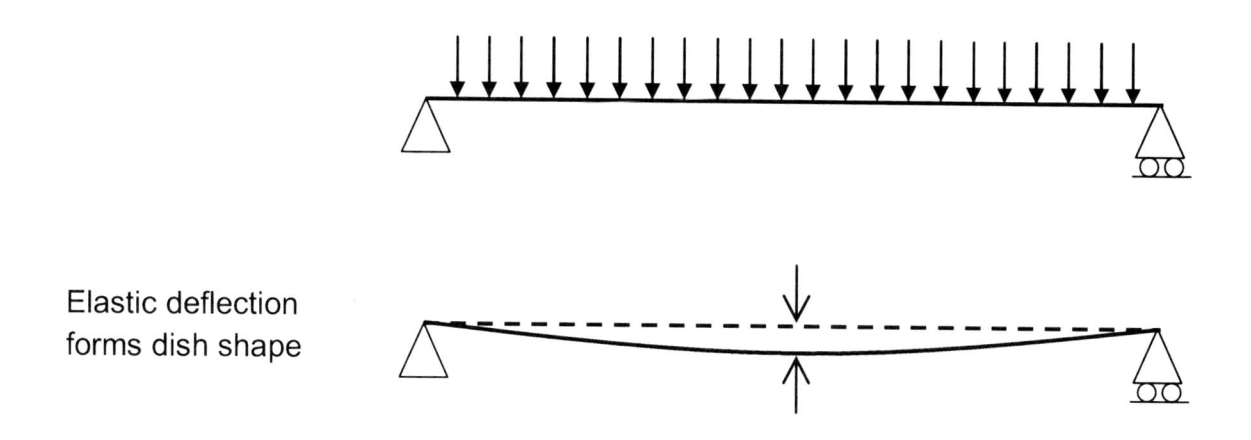

Figure 10.8: Deflection due to ponding

Scupper

50

Undeflected profile of roof

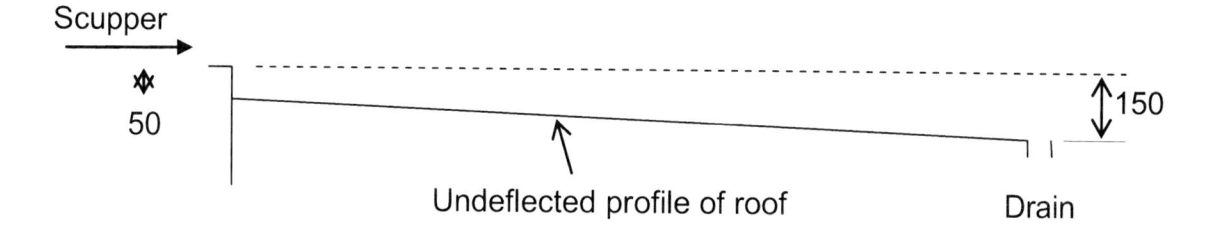

150

Drain

Figure 10.9: Expected profile

Lessons

- The structure was not "damage-tolerant": it had poor resistance to progressive collapse. It was not designed to allow the loss of one hanger.

- Although ponding is mentioned, EC3 does not include a check for stiffness to protect against ponding. This stiffness requirement is often more severe than the conventional deflection limit of span/360 for "brittle finishes".

- Careful of excessive flexibility of long-span roof structures.

- Wind is a dynamic load. It can cause fatigue of members and their connections.

- Do not allow any compressible inserts in connections.

- Effective falls and drains are necessary.

- Incorporate redundancy especially in designing large structures.

Aftermath

The main trusses were undamaged in the collapse and so were reused when the arena roof was rebuilt. The centre of the roof has been raised 760 mm and now slopes towards the perimeter. Fourteen drains have been added on the roof perimeter. The roof is no longer required to act as a temporary reservoir. All hangers were replaced and they were welded to the trusses below. The trusses have been strengthened and the joists made deeper.

The court cases began four years after the collapse and ended in settlement after only two days. The parties shared the $6 million rebuilding cost.

Recommended procedure to anticipate ponding:

Calculate deflections due to DL plus "normal" LL (say 0.75 kN/m^2). If these loads produce a deflection such that the deflected shape acts as a bowl that can store <u>significantly</u> more water, then there may be a ponding problem.

In a tropical climate such as Singapore, the 60 minute rainfall intensity with a return period of 25 years (frequently considered a design value) is 95 mm.

Additional Information

There have been several similar cases of roof failure of steel buildings due to ponding, at least 4 in the US alone, e.g., see Kaminetzky (1991), LePatner (1982).

The American Institute of Steel Construction (AISC) gives formulas to enable the roof stiffness to be checked so that ponding is never a problem. These are incorporated into many American steel codes.

According to the AISC code, all roofs of slope less than 2% are required to be checked for ponding of rainwater or snow. Drains are assumed to be blocked for this check.

Just 150 mm of water is sufficient to collapse an average structural steel roof (designed for a LL of 0.75 kN/m^2).

- Assume roof DL is 0.75 kN/m^2 and has a LL capacity of 0.75 kN/m^2 (i.e., designed to cause a stress of about 165 N/mm^2 where mild steel is used).
- So collapse load for this roof = (0.75 + 0.75)x250/165 = 2.25 kN/m^2.
- Thus the additional load above DL to cause collapse = 2.25 − 0.75 = 1.5 kN/m^2
  = 150 mm water.

10.4 Station Square Shopping Centre, Burnaby, British Columbia, Canada, 1988

[Shepherd & Frost (1995), Feld & Carper (1997)]

This was a one-storey building (69 m x 122 m) with rooftop parking. It was a steel framed structure with a concrete roof on trough decking. The roof was used for car parking. The structure used the cantilevered support system. I-beams passed over SHS columns and extend as cantilevers on both sides. The I-beams supported open-web steel joists which in turn supported the composite deck. The columns were 305x305x13 SHS and were 7.3 m tall. The I-beams were 610 mm deep and 133 mm wide.

Collapse

On opening day of the supermarket, it was thronged with about 1,000 people. Many cars were parked on the roof of the supermarket. Fifteen minutes after the people entered the shop a loud bang was heard and water began to flow from a broken pipe. Looking upwards to the roof, it was noticed that the beam-column connection was severely distorted. The I-beam web had buckled to a near-horizontal position, as the top of the column was displaced horizontally by about 600 mm.

The people were evacuated from the shop. Less than 5 minutes after, a 4-bay portion of the roof collapsed along with 20 cars. The collapsed area was 26.5 m x 22.8 m. Luckily nobody was killed, but 21 people were injured.

Investigation

A panel under a commissioner investigated this failure. The technical cause of this failure was established quickly: there was no lateral support for the compression flange of the I-beam over the column.

Many procedural deficiencies were cited by the commissioner in the project delivery system that allowed such an error to slip through.

- Competitive bidding for design services;
- Unclear assignment of responsibilities;
- Inadequate involvement of designers during the construction phase;
- Poorly monitored changes during construction;
- Incomplete peer-review.

Lessons

- Always remember to brace the compression flange of steel members. A suitable detail would be as shown in Figure 10.10 below.

- Instability is a brittle failure. (However most steel codes incorporate a higher factor of safety to allow for this).

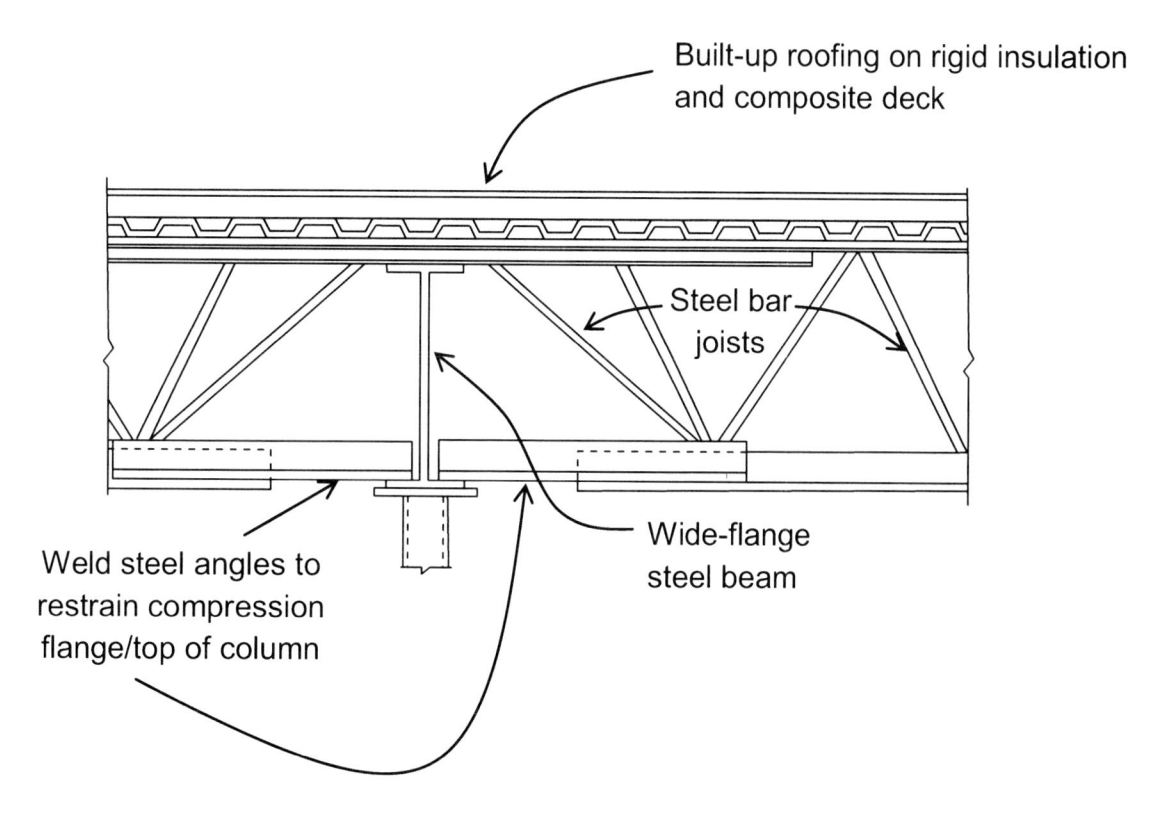

Figure 10.10: Bracing of column top and bottom flange of beam

Additional Information

The structural engineering fee was only 0.3% of the construction cost.

"If you pay peanuts you get monkeys!"

Anonymous

10.5 Koror-Babeldaup Bridge, Palau, 1996

[McDonald et al. (2002), Burgoyne (2006)]

The bridge connected the two Pacific islands of Koror and Babeldaup, part of Republic of Palau, which is about 1,200 km north of New Guinea and 850 km southeast of the Philippines. It was a cast *in situ* post-tensioned box girder bridge. It was designed in 1975 by a world renowned engineer and constructed in 1977. When completed, it was the longest span prestressed concrete cantilever box girder. An elevation is shown in Figure 10.11 and a section in Figure 10.12.

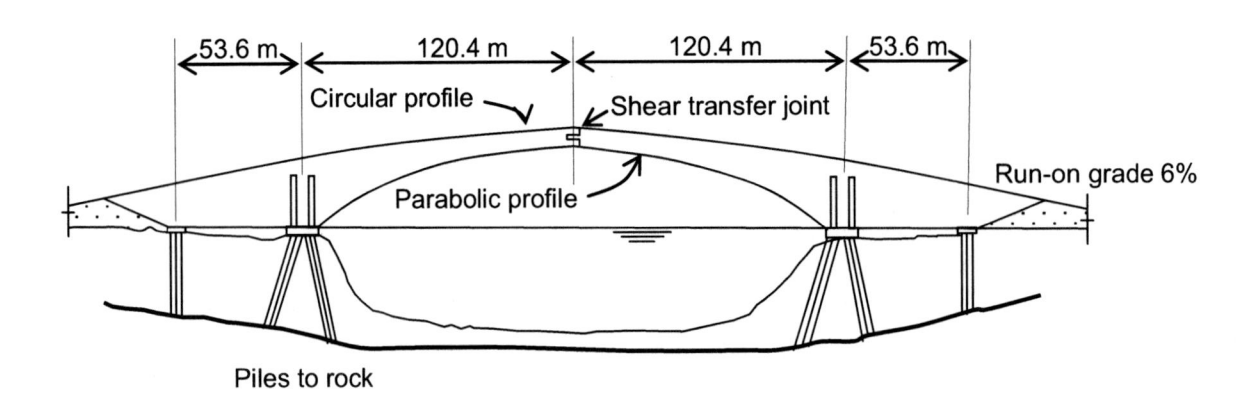

Figure 10.11: Elevation of K-B Bridge Palau

Each half of the bridge was constructed separately starting at the main pier and adding segments on each side symmetrically. The cantilevers had 25 segments each. The post-tensioning was bonded and used 30 mm dia. high strength threaded bars. Over the piers the concrete slab was 430 mm thick and had four layers of PT bars (see Figure 10.13).

Like many other bridges constructed using free cantilever method (Teh 1989), K-B bridge suffered unacceptable time-dependent deflections. By 1996 the deflection at centre of bridge was over 1.2 m. By 1993 three independent investigations by international consulting engineering firms all agreed bridge was safe. One recommended that deflection be cured by the addition of external prestressing and resurfacing. The remedial works to the bridge were intended to restore its serviceability not strengthen it. The work involved elimination of the centre hinge and the addition of external prestressing within the box. The plan was accepted by the client and remedial works to the bridge were done ending in August 1996.

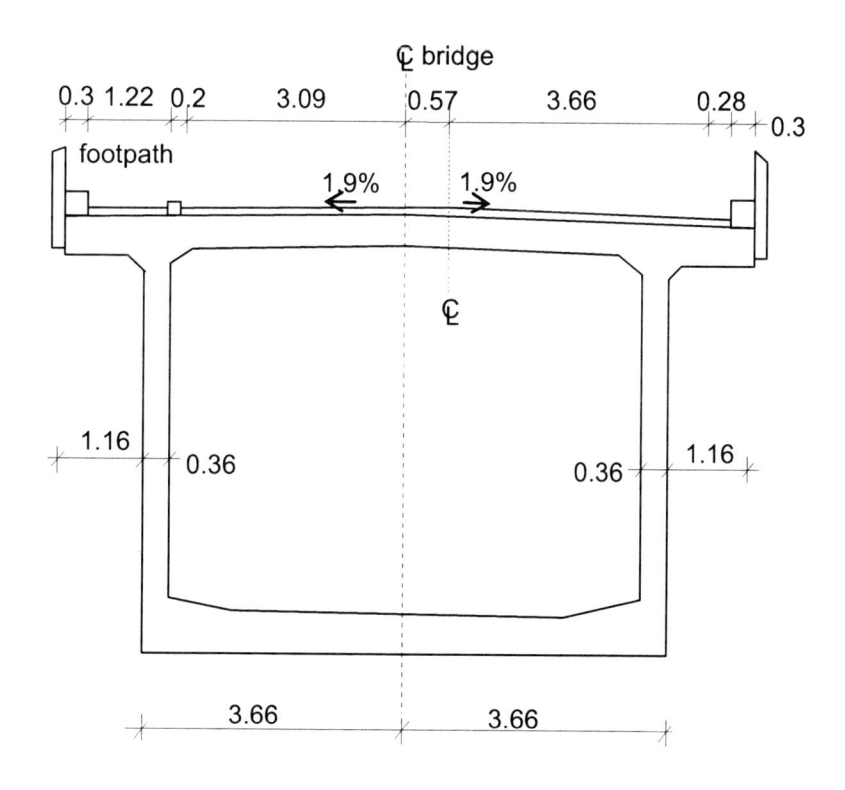

Figure 10.12: Section of Bridge: Depth varied from 3.66 m to 14.17 m

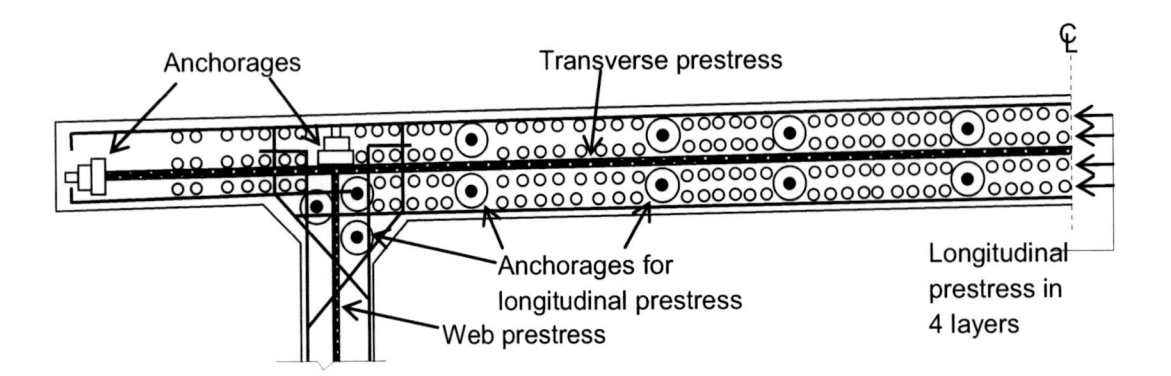

Figure 10.13: Detail of top slab reinforcement

Collapse

On 26 September 1996, the main span of the bridge suddenly collapsed. About 30 minutes before the collapse, breaking sounds could be heard from the bridge. Eyewitnesses described the failure as initiating over pier on the Babeldaup side. The collapse killed 2 and injured 4.

Investigation

Under A-symmetric live load the structure was statically indeterminate to the 3rd degree, but under symmetric load it was statically determinate. This explains fact that there was complete failure once a trigger failure occurred.

The remedial works were investigated and it was found that addition of prestressing etc would not have caused structure any distress.

Inspection of the debris on the Babeldaup side showed that the top slab of the box was badly damaged. Much of top slab's bottom surface had delaminated and spalled, revealing the lowest layer of PT bar. There was clearly much steel congestion. No link reinforcement was present.

The PT bar was not placed uniformly and the horizontal and vertical clear spacing between PT bar ducts varied from 25 mm to 75 mm. The clear spacing was even less at bar splice locations. The fact that PT bar was visible suggested that close spacing of PT bar had resulted in the lower layer forming a weak horizontal plane. (Also, it is likely the concrete in the bottom of the top slab would have been poorly compacted and possibly segregated as there was so much reinforcement above it).

The top surface of the deck slab on both Babeldaup side and Koror side was peppered with pockmarks. Clearly these pockmarks had resulted from the impacts of breakers used in the road resurfacing. The tips of the pockmarks often came within 25 mm of the top layer of reinforcing.

It was found that the Contractor had used 50 kg breaking hammers to do the road surfacing "as the concrete was so hard" (it had a design f_{cu} = 50 N/mm^2). This impact accounted for the delamination of the concrete on the bottom of the top slab. This delamination would have caused the PT bars to lose their bond with concrete and could no longer function compositely with concrete. This would have lowered the shear strength. The result was a shear failure near the main pier on the Babeldaup side. Subsequently the Koror side was overloaded such that the back span lifted until the box compression capacity was exceeded at the main pier resulting in crushing and total collapse of the main span.

Lessons

- If using free cantilever construction method avoid large creep deflections at cantilever tip by making the joint monolithic (usually done since 1970s).

- Congestion of reinforcement can lead to the formation of weak horizontal planes (especially in absence of links).

- Avoid aggressive demolition of a concrete surface. It may damage the concrete causing matrix microcracks and dislodging aggregates.

- Water-jetting or sand blasting is a far better way to remove the surface concrete without damaging the interior. Alternatively, chipping hammers (< 9 kg) may be used.

Aftermath

A new cable-stayed bridge 411 m long opened in 2001 on the same alignment (re-using the foundations).

10.6 Millennium Bridge, London, UK, 2000.

[Dallard et al. (2001), Petroski (2004), Petroski (2006)]

Pedestrian bridges are typically narrow and shallow. Thus they are more easily excited by dynamic forces and have lower damping than road bridges. In 1998 a well-known architect, consulting engineering firm and sculptor teamed up to produce the Millennium Bridge across the Thames from the Tate Modern art gallery to St. Paul's cathedral. It is a 320 m three span steel and aluminium suspension bridge with concrete piers. The sag in the cables is only 2.3 m over 144 m (the main span), making it about 6 times as shallow as a typical suspension bridge. The elevation is shown in Figure 10.14.

Incident

On opening day in June 2000 the footbridge was thronged with people. The bridge swayed from side to side so much that it was closed within two days. It was a major embarrassment as it was felt that at the design stage the bridge had been thoroughly tested in a wind-tunnel and water tank.

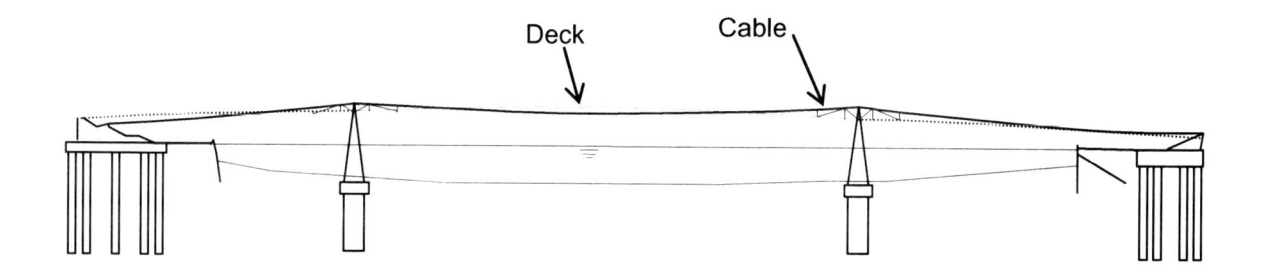

Figure 10.14: Elevation of the Millennium Bridge

Investigation

Research prompted by the behaviour of the Millennium Bridge (there was none before) showed bridges of <u>all types</u> had been affected in past. Thus the problem had nothing to do with the form of the bridge. It was found that in all of these cases, the presence of pedestrians had caused the bridges to swing sideways.

- 1975: Auckland Harbour Bridge.

- 1987: Golden Gate Bridge (pedestrians thronged the bridge on 50th anniversary).

- 1980s: a cable-stayed bridge in Tokyo.

- 1978: A 45 m, 3-span steel footbridge in the "NEC" exhibition centre, Birmingham, UK.

- 1977: Queens Park Bridge, Chester, UK. (80-year old)

- 1999: Pont du Solferino, Paris (New arch footbridge).

- 2000: A firework display on a 100-year old cantilever truss bridge in Ottawa, Canada.

Why did the problems experienced at other bridges not alert the consulting engineer? Because these embarrassments were "buried", e.g., authorities in Paris who closed the Pont du Solferino said the problems were due to the "slippery surface". The problem discovered at the Millennium Bridge could affect *any* long-span footbridge.

Testing by the consulting engineer involved crowds of employees walking across the bridge. It was found that 500-600 people were enough to trigger sway.

When people walk it is well known that there is a pulsating horizontal load applied. This load is applied at a frequency equal to about half that of the vertical load (the latter is applied at about 2 Hz). The horizontal force is about 25 N per person at 1 Hz. The surprise was that everybody on the bridge walked in step and so applied this force at <u>exactly</u> the same time. Previously it was thought that all the sideways forces would be applied at random and so they would cancel each other out.

Subsequent research showed that people will tend to walk in step if the bridge is crowded. When the walkway begins to sway laterally, for whatever reason, (at its own natural frequency) people respond by modifying their walking pace to the vibration of the bridge (as this is easier than fighting the sway). In addition, they place their feet further apart in order to help them to keep their balance. This increases the lateral force.

Thus people walk in step so the pulsating laterally applied load is from <u>all</u> the people. If the lateral natural frequency of the bridge coincides with this frequency then the bridge will sway horizontally in resonance.

Rather than merely being a failure to apply our knowledge, this was a true "state-of-the-art" failure.

Lessons

- It is important that failures are reported fully so that others may learn from the mistakes made.

- It is recommended that the lateral natural frequency be no less than 1.2-1.3 hz if the bridge is to be used by a large number of pedestrians.

- This applies regardless of the form of the bridge. Any long span bridge is vulnerable.

Aftermath

The original cost of the bridge was £18.2 million. At a cost of £5 million, X-braces, 50 tuned mass dampers and 37 viscous dampers were added beneath the walkway. The consulting engineer had to pay £0.25 million. The bridge's appearance was not affected and it reopened in 2002.

10.7 I-35W Bridge, Minneapolis, USA, 2007.

[Hampton (2007), Delatte (2008), Hunter (2008), Ichniowski (2008), Petroski (2009)]

The bridge was a riveted steel truss bridge with a concrete deck. The truss was in the form of a long span over the river, with overhangs at each end. Thus it was statically determinate. It consisted of 2 parallel trusses. The road deck was RC and 8 lanes wide. It was built in 1967 and had a 50 year design life.

Collapse

During a rush-hour on 1 August, 2007 (40 years after construction) the bridge suddenly and completely collapsed. Both the main trusses fell completely (i.e., the main span and two side spans), killing 13 people.

Investigation

A construction crew was carrying out routine "non-structural" maintenance on the South side of the bridge at the time of the collapse. This involved the use of demolition equipment (22 kg breakers) and the storage of materials to be used in the remedial works. They described how the bridge "wobbled" as concrete was removed in the days before the collapse. One lane of the bridge was open in each direction while the maintenance was done.

The bridge had been inspected each year since 1993. (US regulations require inspection every 2 years). The most recent inspection was 2006. Many studies were done on the bridge (e.g., by University of Minnesota in 2001) as it was known to be especially vulnerable (as it was statically determinate). None raised cause for concern. It was given the non-dangerous rating of "structurally deficient" by inspectors in 2005.

Small fatigue cracks were found in the approach spans in 1997: the ends of the cracks were drilled to stop propagation (rounding the hole reduces the stress concentration). No cracks were found on the main trusses. The bridge was considered safe, but was scheduled for replacement in 2020. Corrosion was identified as a risk: Minnesota is cold in winter, and so there was much use of de-icing salts. Corrosion was worsened by bird droppings and mist from a near-by waterfall.

Each riveted connection between the truss members used two gusset plates (one at the front of the members and one at the back). It was discovered that 16 tension gusset plates were grossly undersized (they were 12 mm thick instead of 25 mm thick). The undersized gusset plates were at 8 nodes in the main span. These plates (16) were found fractured in the debris while the remaining gusset plates (208) were unfractured. See Figure 10.15.

Once the defective plate thicknesses were incorporated into the bridge, there was little chance the defect could be detected. Inspection would only ensure bridge did not deteriorate further. Only a re-appraisal would detect the actual defect. Subsequent additions of surfacing over the years only made the problem worse.

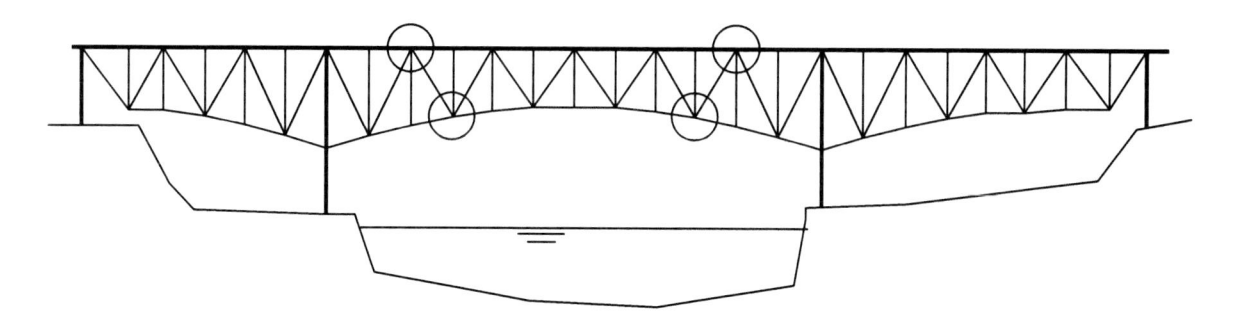

Figure 10.15: Fracture of plates at four locations (2x8 = 16 plates) at the ends of tension diagonals.

The trigger was the unusual loading (an estimated 300 tonnes of construction equipment and material placed on the closed lanes). Unfortunately much of this load was concentrated over the weak gussets. Possibly there was another minor contributory factor: seized bearings as a result of corrosion. This would introduce additional compression stress as bridge tried to expand (the day was hot).

Lessons

- Defects built-in during construction (so called "latent defects") may take decades to appear: triggered by a never-before-experienced load.

- A full reappraisal of structures, especially old statically determinate bridges, is necessary: merely inspecting is not enough.

"This is a classic example of how...a single failure can lead to a collapse. At the time [the bridge was built], it was considered an acceptable risk. Now we try to be more careful."

Spiro Pollalis (bridge designer and Harvard lecturer).

Additional Information

In this case additional load may have helped trigger the collapse. One hundred and sixty years before, in 1847, an additional layer of 125 mm of gravel ballast may have triggered a disaster in: the collapse of the Dee Bridge (designed by Robert Stevenson). The bridge suddenly collapsed under the weight of a passing train shortly after ballast was added, killing 5.

10.8   D. L. Lawrence Convention Centre, Pittsburgh, USA, 2007.

[Post (2007), WJE (2008), Delatte (2008)]

The convention centre is a four storey building approximately 290 m (North) x 182 m (South). The floor structure consists of precast double tee beams supported on structural steel I-beams. It was divided into two by a single expansion joint running North-South (along Grid X9), splitting the building into two 145 m lengths. The structure opened in 2003. The loading bay, where the failure took place, is at South end.

Collapse

The outside temperature was -14°C. The loading bay structure was at a similar temperature. A large truck stopped in the loading bay adjacent to the expansion joint. Suddenly and without warning, there was a collapse of a 10 m x 20 m portion of the 2nd storey floor beneath the truck.

The collapse was captured on CCTV and it showed that the end of the floor adjacent to the expansion joint failed first. There were no injuries. The collapsed area is indicated in Figure 10.16.

Investigation

The original design called for steel seats under the steel beams, but it was later changed to a slotted-hole detail. The detail adopted is shown in Figure 10.17. The expansion joint should have been fully open as the structure contracted to each side. The debris showed that failure had taken place in the weld to the supporting truss; distortion of the angle cleats suggested large axial (tensile) load had been on the steel beam.

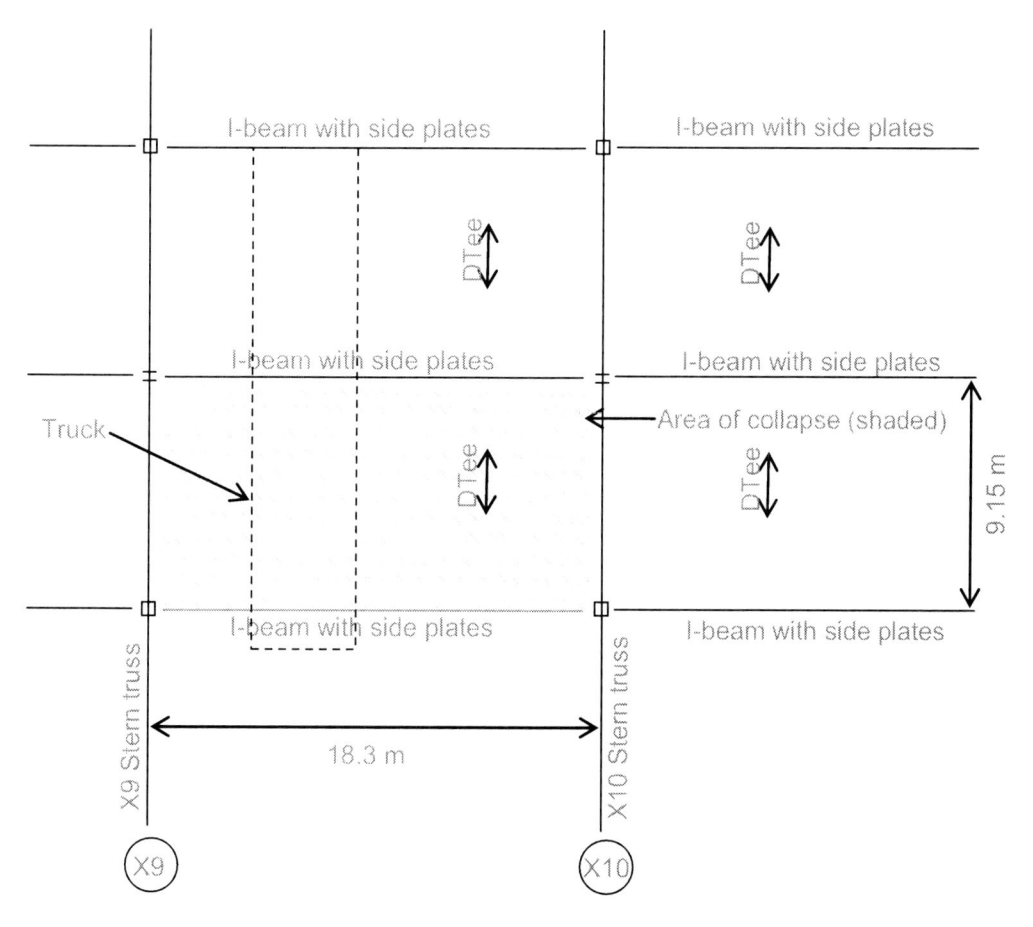

Figure 10.16: Collapsed area

Calculation showed the vertical load at time of failure was 432 kN (increased by about 10% as a result of the truck). No horizontal load was expected or designed-for, but Finite Element Analysis indicated that angle distortion required a force of about 624 kN. The connection failed under these combined loads.

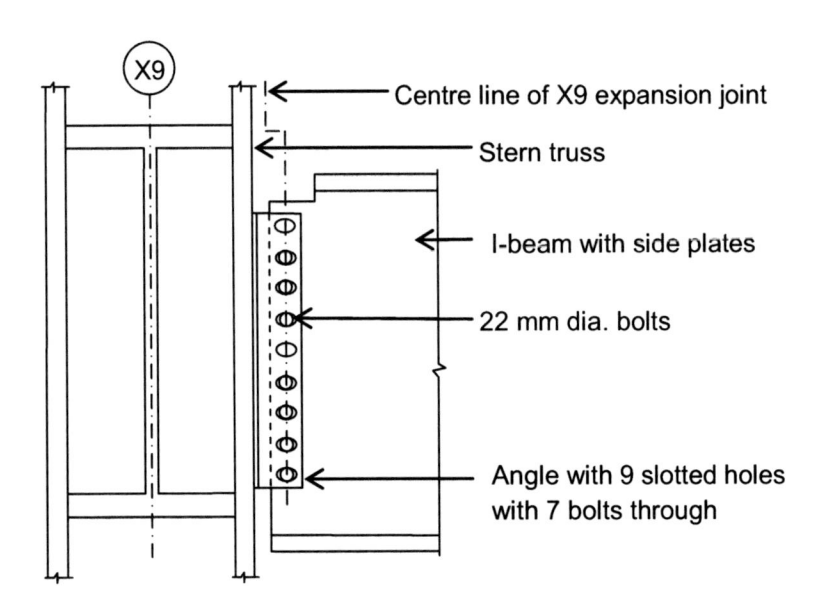

Figure 10.17: Detail at expansion joint

The specification required the use of a steel with a yield point of f_y = 250 N/mm². Instead steel with f_y = 440 N/mm² was used. Thus the steel supported a larger force before yielding. A large horizontal force would only happen if the slotted-hole locked. Unfortunately, the slotted hole was poorly fabricated; apparently the fabricator drilled two holes at each end and removed the material in between leaving a "bump". This is shown in Figure 10.18 (a) and (b).

In addition to lock-up, the vertical load would increase the inevitable bolt bearing and friction. (The extra vertical load of the truck was the trigger for the collapse). Thus the detail, even if fabricated correctly, was unsuitable for large vertical loads.

(a) Correct

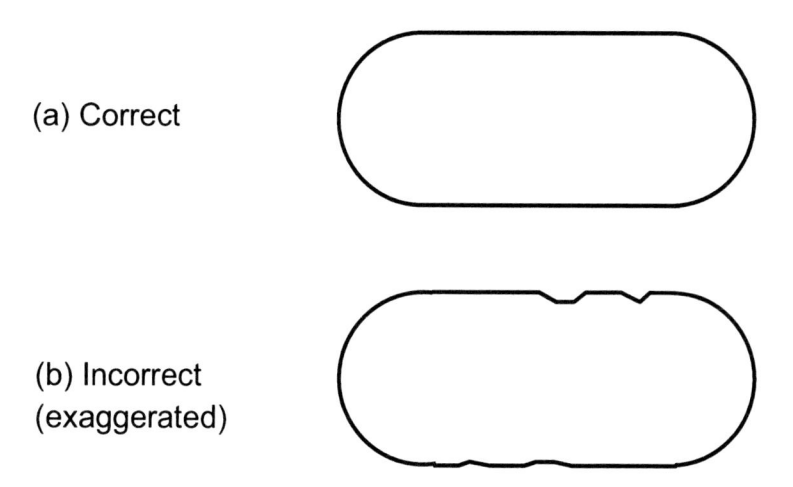

(b) Incorrect
(exaggerated)

Figure 10.18: Fabrication of slotted hole (plan)

Lessons

- The slotted hole connection is a poor choice for a connection, as it depends on:
    - Low vertical load.
    - Good fabrication (to ensure surfaces smooth).
    - Good installation (to ensure slotted holes in correct direction and not skewed).
    - Good corrosion protection.
    - Bolts must be installed at or near middle of slotted-hole.
    - Un-torqued bolts to ensure little friction between the bolt head and the steel around the hole (construction workers usually do not know the purpose of the joint so torque the bolts for "extra strength").
    - Bolt threads should not be on the bearing surface.

- If the steel has too high a yield strength, additional load can be imposed unexpectedly. Here the axial force in the beam could never have reached 624 kN if the yield strength was lower.

- Loading bays are areas of the building likely to be more exposed to outside temperatures.

"I've only seen the slotted hole connection used one other place in an expansion joint in 30 years of doing engineering. And it fell in that place, as well."

Gregory Luth (engineer who worked for firm who designed Hyatt Hotel, Kansas; now owner of successful firm in California; "that place" refers to the collapse of the roof of the steel atrium roof during the construction of the Hyatt Hotel in 1979 two years before the walkways failed) quoted in Delatte (2008).

Aftermath

Repair was done by installing steel seats under the bottom flange of each of the 50 steel I-beams perpendicular to the expansion joint.

Additional Information

In temperate climates it is suggested by EC2 that exposed structures have expansion joints at about 30 m centres. However the low temperature range climate of tropical countries like Singapore, it is recommended that there are expansion joints dividing structure into lengths of no more than 100 m (National Annex of EC2).

Chapter 11: Conclusions

"Redundant systems have also been known to cure severe cases of insomnia"

Russel Fling (former president of ACI)

Observation

The majority of serious failures that take place after construction (over 60% of the cases examined) are due to poor design! The design incorporates a "latent defect" (Reason, 1997) into the structure. Often the defect becomes apparent during or soon after construction. Occasionally, the defect only becomes apparent many years after construction, e.g. I-35W Bridge and Silver Bridge (both 40 years).

These are the state-of-the art failures that have been presented:

- 1972: Service, ultimate, and cladding problems at John Hancock Tower, Boston.
- 1994: Brittle failure of welding during Northridge Earthquake, California.
- 2000: Service problems at Millennium Bridge, London.

These all relate to the gradual increase in our understanding. Thus the vast majority of failures are not state-of-the-art: i.e., they could have been prevented (or at least mitigated) by the application of our existing knowledge.

By improving the quality of our designs, we not only address these "latent defect" problems directly, but we can also mitigate the failures that happen for any other reason.

Summary of pitfalls

- **Poor Quality Control/checking**, e.g., Hyatt walkways, Compassvale School, Hotel New World, Sleipner.

- **Poor Redundancy**, e.g., Ronan Point, Bailey's Crossroads, Hartford, Kemper, Alfred Murrah Building, Cocoa Beach, Willow Island.

- **Extrapolation based on a successful structure**, e.g., Tacoma Narrows Bridge, Use of restrained brickwork on tall buildings.

- **Corruption**, e.g., Sampoong, Hotel New World, Compassvale School.

- **Misinterpreted/Inadequate wind tunnel tests**, e.g., Tacoma, Ferrybridge, Hancock, Citicorp.

- **Failure to maintain/inspect adequately**, e.g., Piper's Row, Silver Bridge, Mianus Bridge, Schoharie Creek Bridge, Hotel New World, Alexander Kielland, Gwas Bridge, Paris Airport.

- **Work outside your area of expertise**, e.g., Cocoa Beach condominium.

- **Poor Modelling**, e.g., Sleipner, Hartford, Compassvale School, Ramsgate Walkway.

- **Extrapolation from small-scale tests**, e.g., Quebec Bridge, Northridge steel.

- **Extrapolation from short-term tests**, e.g., Stepney School.

How much extra should we spend to reduce failures?

Safety on the railways provides us with an example. The UKs Health and Safety Executive recommend a cost-benefit analysis to decide if a safety improvement is economically justified. That such an exercise is controversial when dealing with human life is well known. However, they suggest that spending of at least £1M is justified for a human life. Clearly the additional cost is likely to meet this limit. The figure used by the US Federal Aviation Administration is $2.7 million (Bazerman & Watkins 2008).

Eg., Typical R.C. frame:

Cost of Reinforcement approximately equivalent to => 900 tonnes of extra reinforcement to save a life. Thus the extra expense is usually justified!

Repeated Failures

- Failures due to poor detailing of RC, e.g., Sampoong, Baileys Crossroads, Piper's Row, Sleipner, De la Concorde Overpass.

- Unnecessary use of large statically determinate structures, e.g., Hyatt Walkways, Kemper Roof, Schoharie Creek Bridge, Mianus Bridge, I-35W Bridge.

- Structures where maintenance was unnecessarily difficult, e.g., Mianus Bridge, Silver Bridge, and De la Concorde overpass.

- Restrained brickwork cladding and stone cladding failures, e.g. Singapore Condominium, and Amoco.

- Ponding, e.g. Kemper.

Danger Signals

- Lack of *independent* oversight of design and construction.

- Poor communications between parties.

- Chain of responsibility unclear.

- Overconfidence.

- Excessive Time/Fee pressure.

- Structures that have any of these features (The Institution of Structural Engineers 2002):

    - have minimal redundancy.

    - barely meet code.

    - are likely to attract many people.

- use innovative design or materials.
- exist in an aggressive environment.
- were designed to now outdated codes or fall outside scope of code methodologies.

Shortcomings of Code Approach (EC2 (BS 2005a) **&EC3** (BS 2005b))

- No requirement to vary factor of safety for the importance of the member (consequences of failure).
- Concrete shear provisions inadequate (EC2):
    - By adopting a similar factor of safety for flexure and shear, the code effectively assumes that shear failure is not likely to trigger a progressive collapse. This is only true if the member is not statically determinate and has sufficient bottom reinforcement;
    - The member designation "slab" may not need shear reinforcement but a "beam" usually does;
    - If the axial compression is large the code may indicate incorrect shear reinforcement.
- Safety factors are more appropriate to failure of statically indeterminate structures than statically determinate structures.
- The factors of safety for the various types of brittle failures are inconsistent. Logically the FS against brittle behavior should be greater than that against ductile failures. (EC2). Any underestimate of flexural capacity (e.g., steel or concrete of higher than characteristic strength) further increases the likelihood of shear failure. (EC2). The approach to shear design recommended is instead, to use the estimate of flexural overstrength to derive the shear to be used for design, i.e., similar to the approach used in earthquake design.

- Approach to progressive collapse resistance by merely requiring ties is often inadequate, since the effect can be to increase connection loads so more of a structure is dragged down. Similarly, there is no acknowledgement progressive collapse can, in some circumstances, be increased by **decreasing** continuity (i.e., increasing isolation). The latter solution is more appropriate to bridges than to buildings as noted by Starossek (2009) due to the importance of not allowing a floor of a multistory building to fall and so initiate a progressive collapse.

- Failure of some elements, e.g., RC slabs by punching or RC tied columns, is brittle and so should attract a higher factor of safety.

- Not required that consideration be given to thermal expansion of members during a fire.

- Overall factor of safety for low/no LL structures, e.g., water-retaining structures, is too low.

- Length and diameter of bottom bars not controlled (EC2)

Note that EC3 like most other steel codes includes an extra factor of safety when buckling is a risk, bringing the overall safety factor to about 2 (Heyman 2008).

Why do the provisions of codes matter? What about "Engineering Judgment"?

- Most catastrophic failures are design related.
- More design being done by contractors, and so strictly according to code.
- Computer programs implement code equations exactly.
- More inexperienced engineers (to cut costs).
- More "Value Engineering" exercises.
- Consequences of out-of-date codes with us for a long time.

"Adhering to accepted practices and observing standards is not sufficient. Such behaviour is called minimal compliance. *It neither guarantees a safe product nor an excuse in court."*

Martin and Schinzinger (authors of *Ethics in Engineering, 1997*)

Lessons for Design

1. The Factor of Safety needs to be given more careful consideration. A **general** increase in the Factor of Safety is unnecessary and inefficient. A higher Factor of Safety is to be used for brittle modes so that as the load is increased from zero, a ductile mode is triggered first. Use an appropriate Factor of Safety applied to failure type in **targeted** manner (recommended in Table 11.1 below):

Table 11.1

		Consequences of failure	
		Low	High
Ductile	SD	2.0	2.5
	SI	1.6	2.0
Brittle	SD	2.5	3.0
	SI	1.8	2.5

Notes to Table:

- "SI" => design ensures initial failure at one location only serves as **warning** of total collapse.

- "SD" => little or no provision for load sharing, so that collapse of major portion of structure results from collapse of this element. Note that a statically determinate structure with provision for alternative load paths in the event of failure requires only the "SI" safety factor. In addition, a statically determinate structure with "collapse resistance" (e.g., designed for a high load determined by a risk analysis, say 34 kPa as required by EC2) qualifies for the "SI" factor even if no alternative paths are provided.

- "low" consequences of failure: where access to the public is not allowed, e.g., private dwellings. Consequences considered "high" otherwise.

Example 1: RC continuous beam with cantilever overhang

- Design load for design for flexure:
    - Ductile failure; statically indeterminate (SI);
    - FS = 1.6-2 for low or high consequences

- Interior of beam:
  Design load for design for shear:
    - Ductile failure (as alternative load paths); statically indeterminate (SI);
    - FS = 1.6-2.0 for low or high consequences

- Exterior cantilevers:

  Design load for design for shear:
    - Brittle; statically determinate (SD);
    - FS = 2.5-3.0 for low or high consequences

Compare with EC2: FS = 1.15x1.35 = 1.55 and 1.15x1.5 = 1.725 for DL and LL respectively; average 1.64.

Example 2: RC tied column in building

- Failure is brittle; Statically indeterminate (alternative load-paths available);
- "Low" consequences => FS = 1.8
- "High consequences => FS = 2.5

Compare EC2 value of 1.35x1.5 = 2.0 or 1.5x1.5 = 2.3 for DL and LL respectively (taking concrete only). Average = 2.2

Example 3: Simply supported steel beam

- Beam designed as statically determinate.
- However provided bolts at each end to allow catenary action once the steel beam has failed due to excessive moment. Thus it qualifies for the "SI" factor.
- Thus Safety Factor recommended is 1.6-2.0 depending on consequences and assuming lateral buckling is not possible so failure ductile.

Example 4: RC tied Tee column

- Statically determinate; shear is brittle => 2.5-3.0, while flexure is ductile => 2.0-2.5

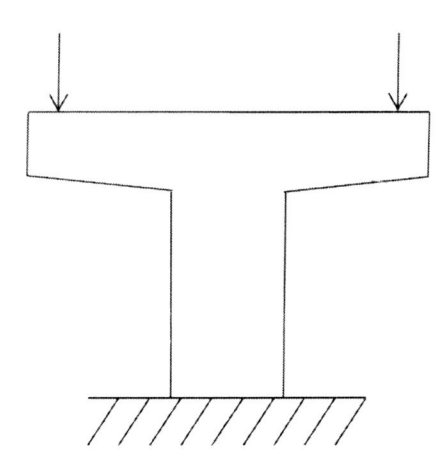

Other cases

- Similarly a timber floor, even if statically determinate, allows load-sharing so qualifies for the "SI" factor.

- Also, a simply supported precast concrete beam with tie steel in the topping allows catenary action so qualifies for the "SI" factor too.

- If the flat slab connection incorporates bottom reinforcement, then it too qualifies as "Ductile" even though the phenomenon of punching is, like ordinary shear, brittle.

Minimum Factor of Safety?

*"in a capably executed and conventional design, an estimate of the **dead** load supported by a member might be off by 15% to 20%"*

Council of Tall Buildings and Urban Habitat (CTBUH) Monograph on Planning: Vol CL-Tall Buildings: Criteria and Loading (Robertson 1980).

(The higher figure given by CTBUH is likely to apply when the tributary area method of column design is used. Modern computer programs implement the elastic design so there is slightly less uncertainty about the reactions).

Thus to cope with variations in the loading (say 15%) 1.15.

So the lesson 1 concerned the Factor of Safety. The other major lessons are:

2. **Connections** are a common source of trouble (e.g. Hyatt, Sampoong, etc). Design so that the connection is **not** the weak-link, e.g. use a higher factor of safety on connection design, so that the member fails (ductile) before the connection (brittle). Incorporate bottom bars into slab/column connections in concrete.

3. Design so that structure is sufficiently <u>stiff</u> not just sufficiently <u>strong</u>, e.g. P-delta effects, rainwater ponding, restraint to compression element, wind stability problems.

4. Avoid using details which will be difficult to maintain in future, e.g., pin connections in mild steel (instead use stainless carefully isolated to prevent dissimilar metal corrosion), RC halving joints.

5. Know the site! Saving money on the site investigation is a false economy, e.g. in Abbeystead the designers did not realize there was even a risk of methane.

6. Do not take unnecessary risks, e.g. Big-Dig fixings held a concrete panel overhead (and it was well known that epoxy resin has poor temperature resistance and the long-term behaviour is suspect). Simply using a lighter panel would have reduced the risk.

7. Assess the risk properly: e.g. in residential structures there is a higher probability of a gas explosion especially if the gas is piped.

Recommendations for Analysis

- Elastic generally analysis gives reasonable results upon which a design can be based.

- However, **sensitivity** checks are needed to assess how changes in the boundary conditions are likely to affect the design, where the service behaviour influences the ultimate (e.g. P-delta effects, ponding).

- In some design offices an elastic analysis is done and the column axial loads obtained are arbitrarily increased by 10%. In my opinion this is a good practice.

- The shear load should be based on the actual flexural moment provided, with allowance for overstrength.

- Statically indeterminate trusses must be conservatively designed.

- Full-scale testing to assess accuracy of computer models is useful. As well as conventional load tests, parameters like the natural frequency can be measured in-situ (Grossman 1990).

Reducing Failures

Recall there are two aspects to our problem:

- Lowest failure rate possible is not zero.
- Bigger structures ensure consequences of failures worse.

Solution 1: Checking

1. Improved **Quality control**:

    - To catch inevitable mistakes.
    - Thorough, independent, checking of design.
    - Done by appropriately experienced people.
    - Checking of design should be management-driven (otherwise it may not be done properly).

- Adequate site supervision/inspection.
- Adequate and timely maintenance/re-appraisal as new knowledge becomes available.

The need to check (Reason, 1997):

- Human error is inevitable.
- The issue is not why the mistake was made, but why it wasn't detected before construction.
- A vital function of management is to try and catch the errors (slips and errors of judgment).
- However, the structure is usually checked only for code compliance.

Note: Appendix 2 contains a check list the author has used to check analysis for the designs of buildings prepared using ETABS software (from Computers and Structures, Inc). The spreadsheet referred to is based on Robertson's formula and the predictions of various codes. It is still of the utmost importance to check the drawings too.

Limitations

- There are many examples of failures where structure *satisfied* the code current at the time. e.g., Sampoong etc (i.e. punching cases), Hancock, Alfred Murrah, Palau Bridge.
- Thus failure would not have been prevented merely by an increase in quality control.
- So the best checking can do is to keep the failure rate low.

There must be a second part to the solution.

Solution 2: Redundancy

Ensure there is **redundancy** in the structure so that any local failure (for whatever reason) merely serves as a warning, i.e., structure is tolerant of damage. In many cases this involves little or no increase in the cost of the structure.

Increasing Redundancy

- Avoid large statically determinate structures as permanent structure.

- If these must be used, ensure alternative load path in event of failure, e.g., transverse members, cradles.

- Add bottom bars to flat plate structures.

- Ensure ties are provided (as required by code).

- Ensure that all structures are **detailed** to ensure redundant behaviour.

- For **normal** structures (e.g., conventional office or residential buildings) this **detailing** approach is likely to be sufficient (unless there are elements needing special attention e.g. transfer beams).

- For special structures (e.g., large sports arenas or shopping malls, etc) a **design** approach is necessary: explicitly consider the removal of each column member and ensure the frame can remain stable and support say $DL+50\%IL$. Any damage should be limited to not more than about 15% of the floor area.

"the assumption of perfect reliability combined with no redundancy and no just-in-case strategy seems irresponsible and naïve, but is consistent with the way many people manage low probability risks"

Marc Gerstein (Author of *Flirting with Disaster*, 2008).

Education & Checking

- Fragmentation of consulting industry.
    - Means more fee competition and/or pressure to do more jobs faster ("more efficiently").
    - Less investment in education.
    - Less likely to carry out checking.
- These are **FALSE ECONOMIES**.
- The real risk is in not investing in these two.
- Aside from social responsibilities, few organizations can **afford** a structural failure (Reason 1989).

Summary: Reducing failures

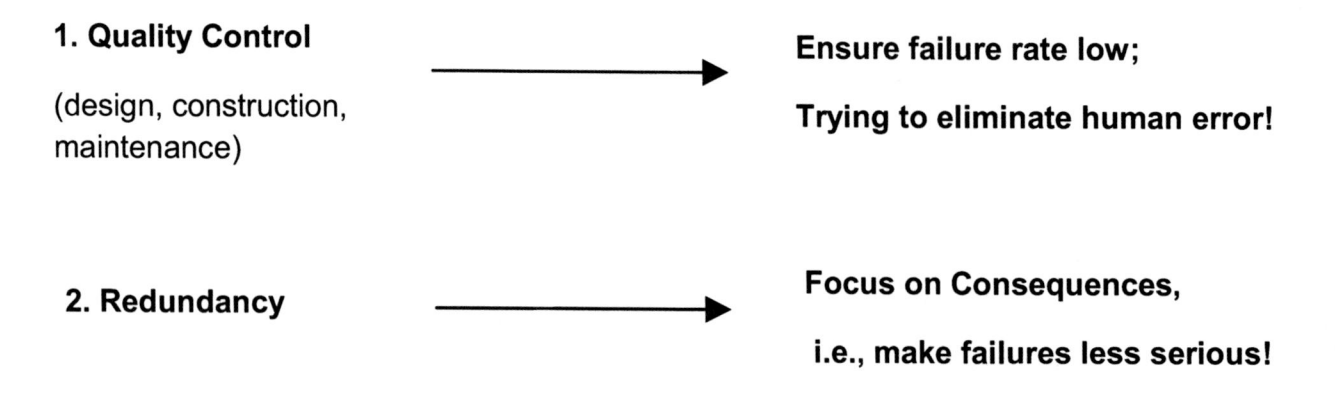

1. Quality Control
(design, construction, maintenance) ⟶ **Ensure failure rate low;**
Trying to eliminate human error!

2. Redundancy ⟶ **Focus on Consequences,**
i.e., make failures less serious!

```
           ┌─────────────────────┐
           │    GOOD DESIGN      │
           └──┬───────────────┬──┘
              │               │
           CHECKING        REDUNDANCY
              │               │
        ////////////////////////////
                EDUCATION
```

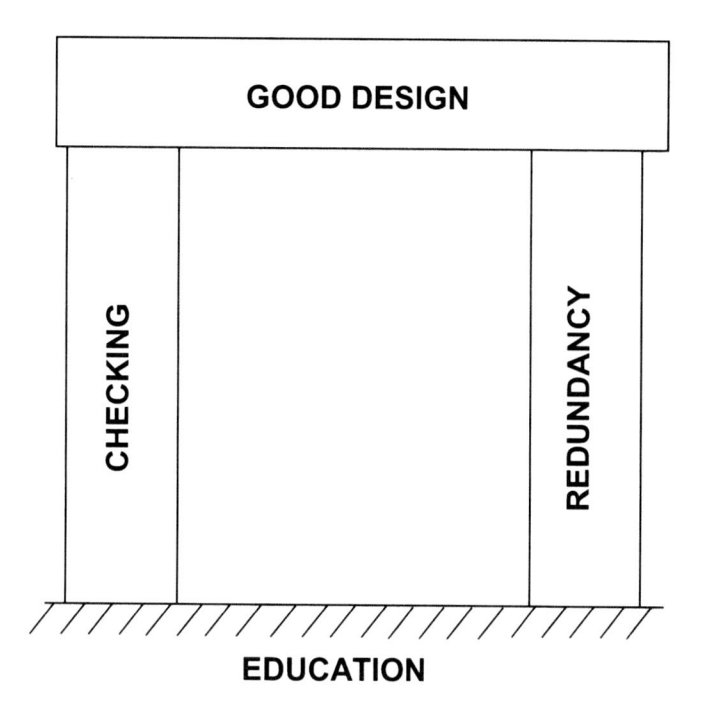

"Only a deeper consciousness of our human and social responsibilities can lead to the construction of safer buildings"

M. Levy and M. Salvadori (Authors of *Why buildings fall down*, 1992)

Worked Examples:

Quiz 1: Statical determinancy and indeterminacy

Demonstrate that the following two structures are statically determinate.

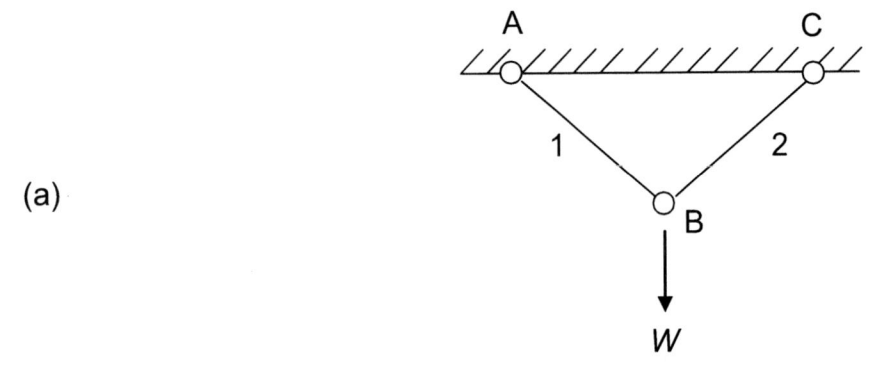

(a)

(b)

Quiz 2: Structural appropriateness 1

(a) Is the following structure suitable as the entrance to a sports stadium?

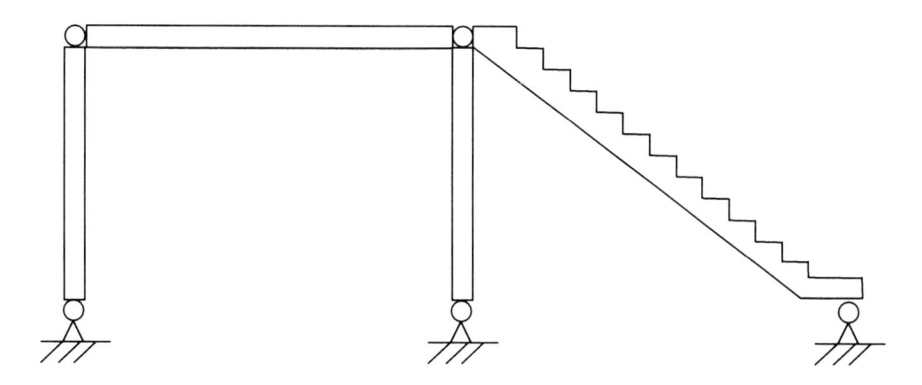

(b) Is the following structure suitable as the entrance canopy over a private residence?

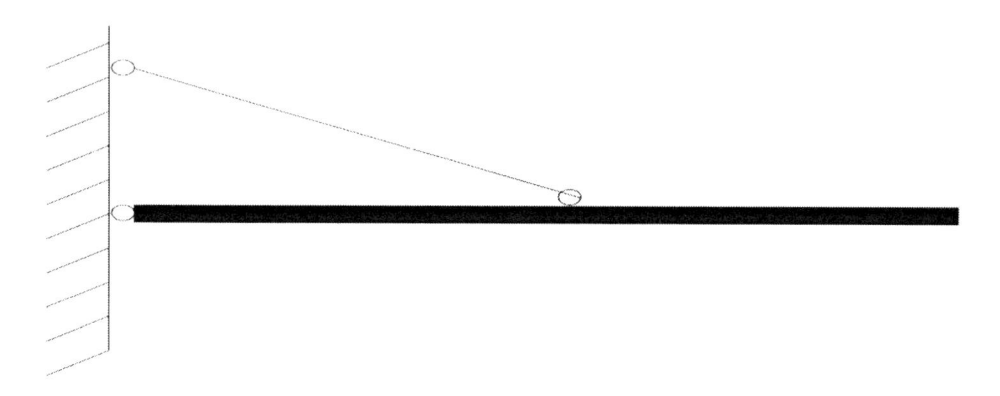

Quiz 3: Structural appropriateness 2

Is the following structure suitable as one of the support trusses for a long-span roof over an auditorium?

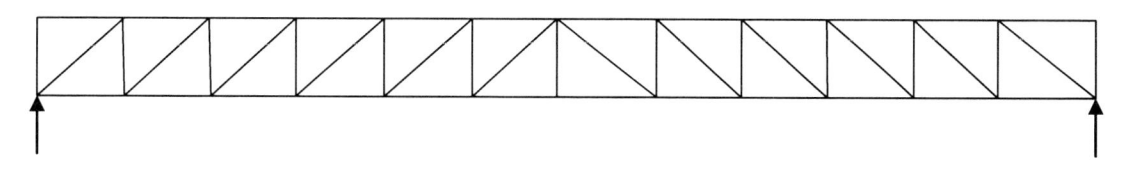

Quiz 4: Checking

When checking computer analysis output, a rough estimate of the critical values is often all that is required.

In the following beam make an estimate of the maximum hogging and sagging bending moments.

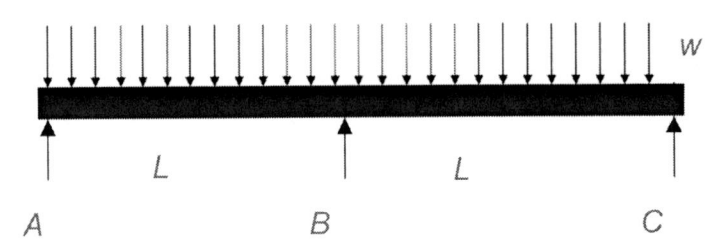

Quiz 5: Probability and Risk

Suppose a stock-exchange trader knows the following will happen if he/she trades:

There is a 999/1000 probability of earning $1 but a 1/1000 probability of losing $10,000. Is it worth the risk to carry out this trade?

Quiz 6: Natural frequency

A steel I-beam 254x146x31kg/m UB is used for the following cantilever. Estimate its natural frequency:

$E = 205,000$ N/mm^2; $I = 44,390,000$ mm^4

10 m

Quiz 7: Improving redundancy

A bridge of two equal spans is to be constructed using simply supported precast prestressed beams and an *in situ* topping.

Show (a) how the *in situ* topping can be used to ensure continuity at support B; (b) how temperature variations may be accommodated; and (c) list the advantages and disadvantages of the scheme.

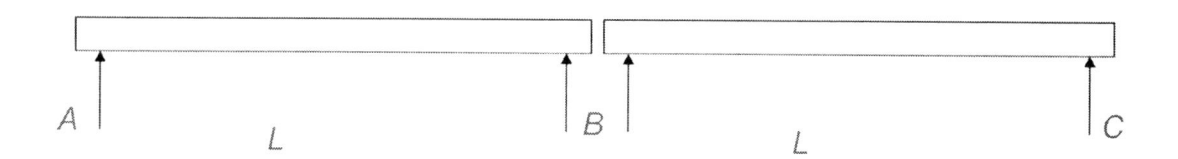

A   L   B   L   C

Quiz 8: Second order effects

The frame shown deflects under the load P.

The original geometry as well as the deflected shape are shown. Write expressions for the first order and the approximate value of the total moments.

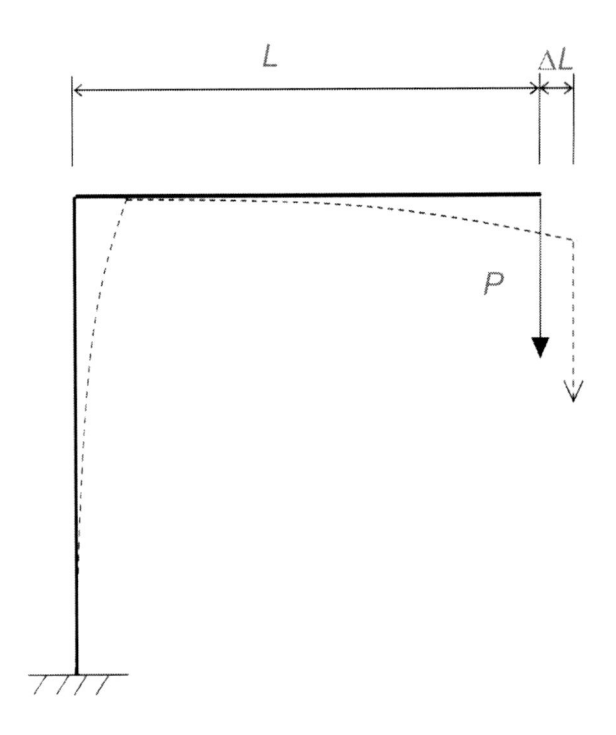

Quiz 9: P-delta effects

The following three storey car park is to be used in a low-wind environment, say in Singapore or Malaysia. Its plan dimensions are 16 m by 32 m. The floors are 250 mm thick flat plates. Neglecting any other weight (i.e., cladding and columns), estimate the **modification factor** in the long direction of the building assuming the frame is designed to move $H/500$ at the roof under a wind load of 0.8 kN/m². Assume the deflected shape is linear and concrete is 24 kN/m³.

32 m

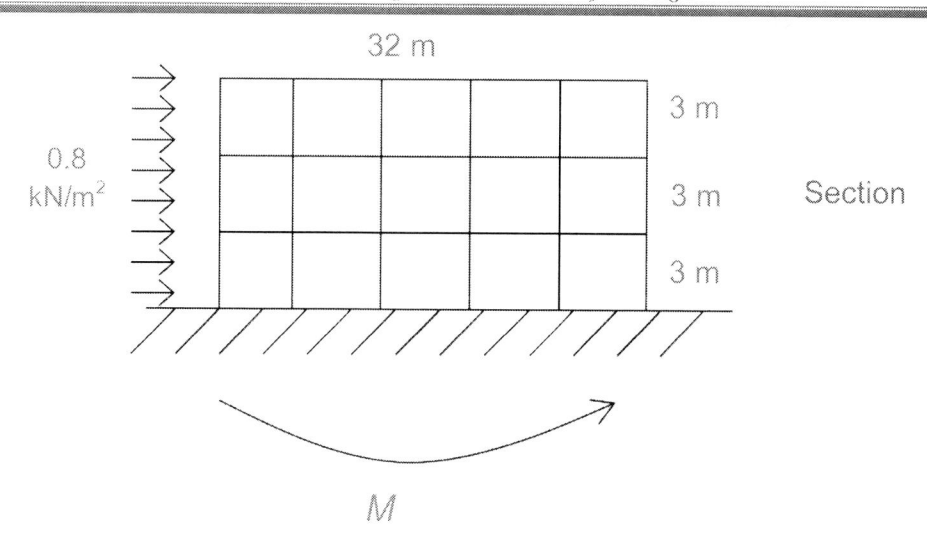

0.8 kN/m²

3 m
3 m
3 m

Section

M

Quiz 10: Marble column

This problem was first described by Galileo as noted in Petroski (1994).

A marble column is stored outdoors resting on two wooden supports placed near the ends.

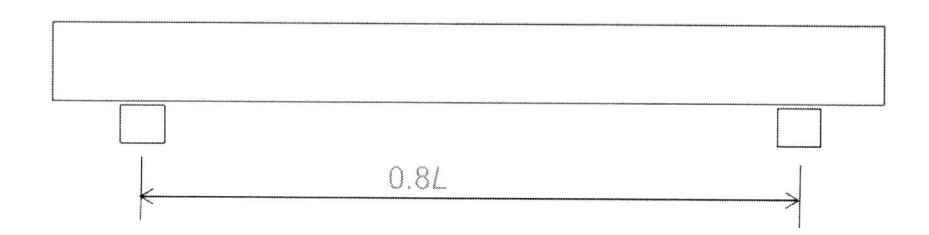

0.8L

Later somebody has the idea of adding an extra wooden support at midspan.

It was agreed by all that this would help prevent the column cracking. The additional wooden support was placed.

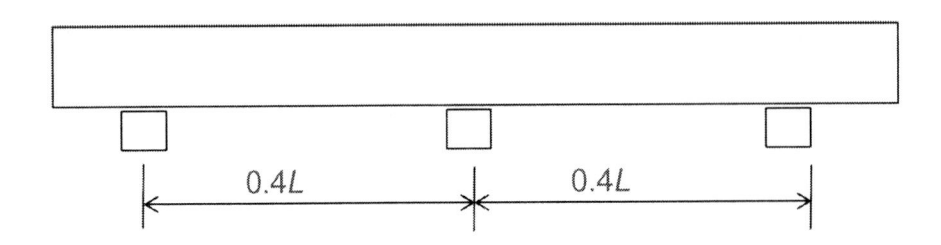

Months later the marble column was found to be cracked, directly over the new support.

What had happened?

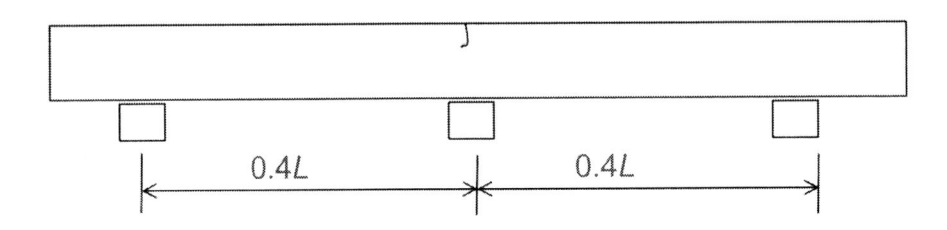

Quiz 11: Exposed long-span trusses

The following structure is to be used as a long span roof structure. How can the roof be designed such that failure of one truss does not result in failure of a large part of the roof? As the roof is exposed to view, no clearly visible out of plane members can be added. Suggest a way the redundancy of the roof could be further improved. Roof trusses are steel and are at 6 m centres.

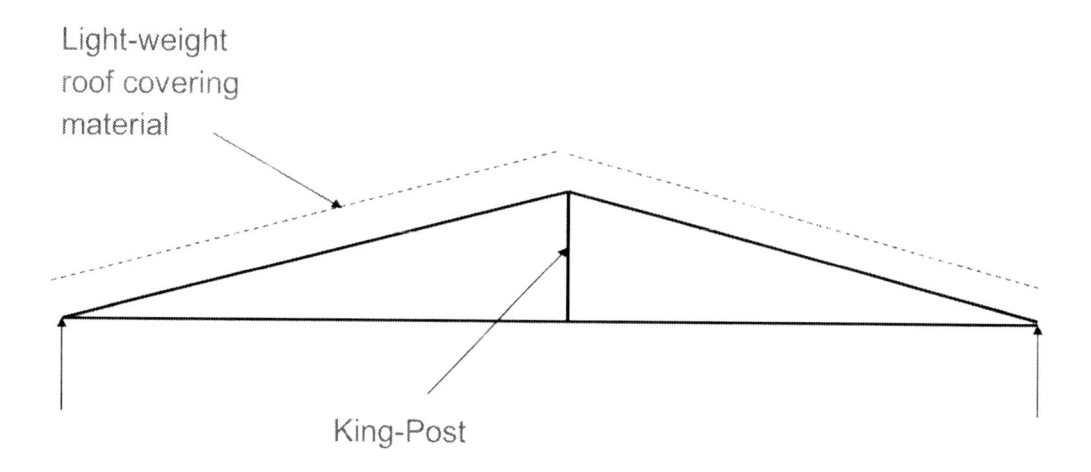

Quiz 12: Redistribution of moments in RC

Consider a fixed ended RC beam supporting a UDL of *w* throughout the span.

Suppose it is decided to *ignore* the fact that it is a <u>fixed ended beam</u> and just design it as if it were a <u>simply supported beam</u> supporting a UDL., i.e. 100% redistribution! What would happen?

Quiz 13: Use of Robertson's Formula for John Hancock building.

Given the following data, calculate the likely value of the modification factor in the long direction of the building:

- Building height = 234 m; plan 90x31 m

- Building density, $W = 190$ kg/m^3
- Design drift ratio at top of building, $R = 1/400$.
- Design wind pressure is 1,400 N/m^2.

Quiz 14: Checking safety of proposed construction cycle

If two levels of propping are to be used, check whether a construction cycle of 7 days is safe given the following design loadings:

- Dead Load (slab) = 200 mm slab = 0.2x25 = 5.0 kN/m^2
- Additional Dead Load = 1.5 kN/m^2
- Imposed Load = 2.5 kN/m^2

Quiz 15: Piper's Row

Estimate the resistance to load of the level 3 slab of Piper's Row car park, level 4 of which collapsed (see case study). Use EC2 load factors.

Quiz 16: Progressive collapse resistance

A 17-storey concrete office building consists of one-way beam and slab construction. Typically the column grid is 6x6 m. On the first suspended level (level 2) each alternate column is transferred by an RC transfer beam. Level 1 consists of a retail area. Risk analysis suggests that the main risk is likely to come from internal explosions and fire.

Assume that all wind is resisted by shear walls. The code (EC2) approach (i.e., vertical ties) is considered inapplicable due to the presence of the transfer beam.

Suggest measures to increase the progressive collapse resistance of the transfer structures and of the upper structure.

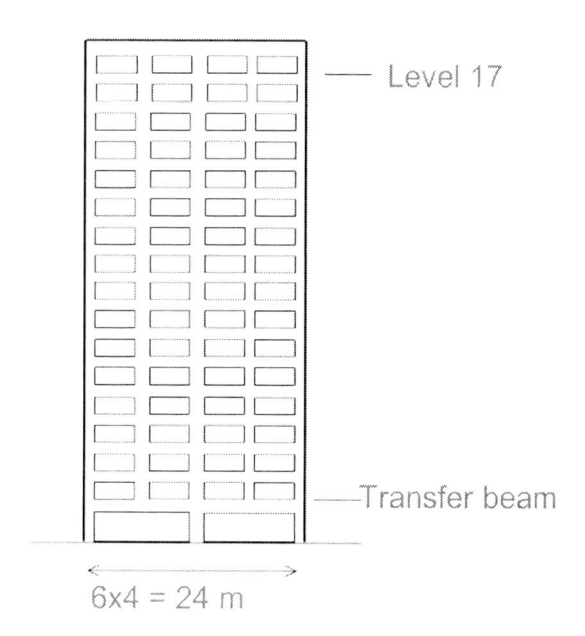

```
                    Level 17


                    Transfer beam
    6x4 = 24 m
```

For this accident case, the ultimate column load is estimated to be 24,000 kN due to DL and permanent LL (i.e. $DL+50\%IL$).

Quiz 17: Safety factor of tied column according to EC2

Consider the column whose cross-section is shown below. Estimate the factor of safety provided by EC2. Assume it is non-slender and that the materials have their average strength. Ignore any bending moment (i.e., assume column is in pure compression). Assume also that the axial load is equally due to dead and imposed load.

300 mm

300 mm

4T25

Characteristic cylinder strength of concrete, f_{ck} = 30 N/mm^2

Characteristic yield strength of steel, f_{yk} = 500 N/mm^2

Quiz 18: Safety factor for shear and flexure according to EC2

Estimate the mean factor of safety in shear and flexure recommended by EC2 for the following beam. Assume the characteristic material strengths are f_{ck} = 40 N/mm^2 for concrete and f_{yk} = 500 N/mm^2 for steel, and mean strengths are f_{cm} = 48 N/mm^2 and 550 N/mm^2 for concrete and steel respectively.

DL = 2.25 kN/m

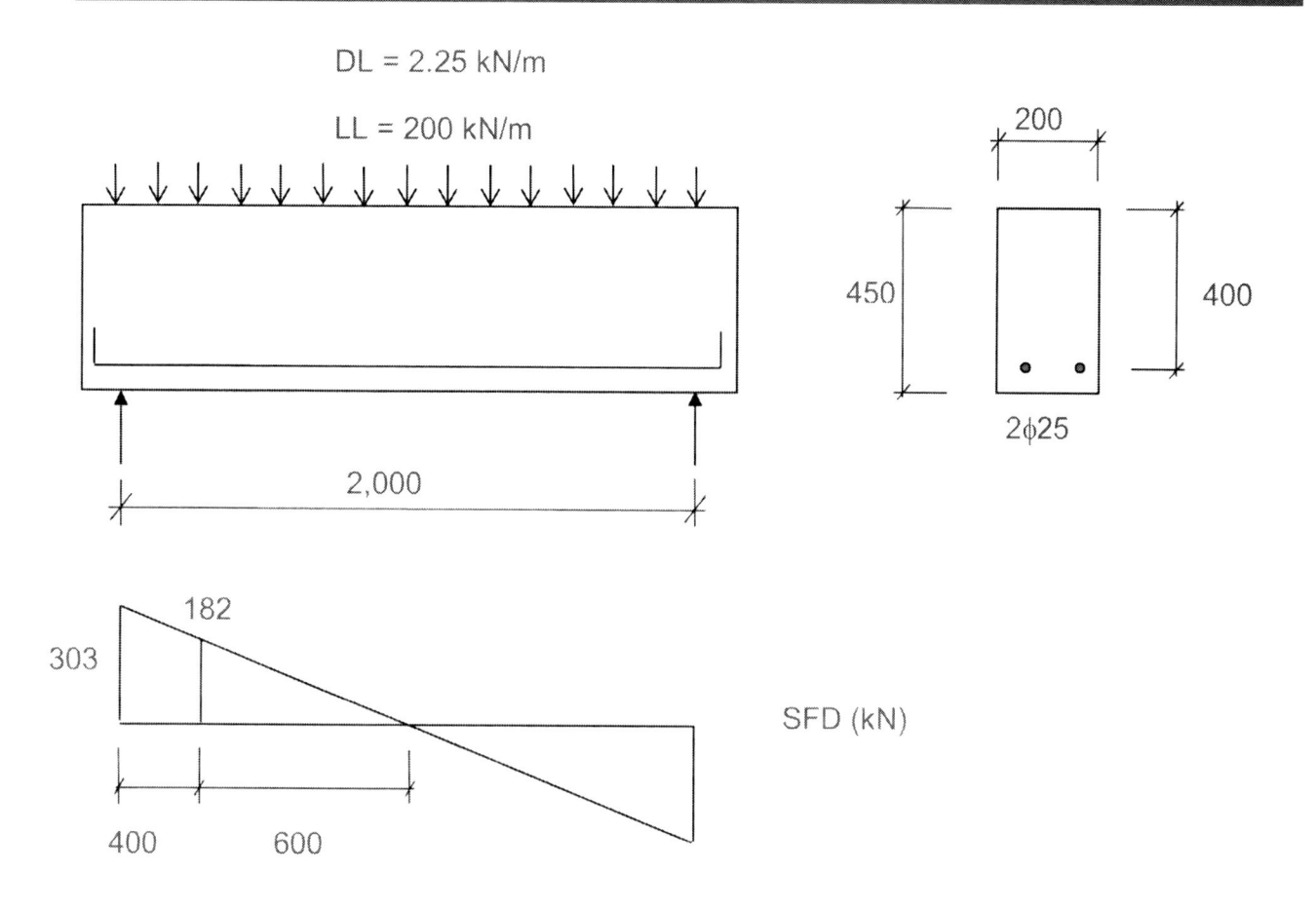

LL = 200 kN/m

Quiz 19: Ponding

A roof is constructed using a 407x178x74 UB (74 kg/m) spanning 15 m simply supported. The steel is Grade 50. The DL is 6 kN/m, LL based on 0.75 kN/m^2 is 4.5 kN/m. Second moment of area, I is 273x10^6 mm^4. Check if ponding is likely to be a problem.

Quiz 20: Restrained brickwork

Suppose a brick wall of 100 mm thick is 3,000 mm high and is built tightly to the underside of the concrete slab.

An insitu stress-relief test shows the wall is resisting an axial compression of 3 N/mm². Is this an excessive stress? Assume the clay brick units have a characteristic strength of 50 N/mm² and the wall is built with a general purpose mortar (e.g., cement:lime:sand of 1:1:5).

Quiz 21: Bottom bars in flat plate structures

A flat plate structure has spans of 6 m in each direction. It is 200 mm thick. Assume SDL = 1 kN/m² and a live load of 2.5 kN/m². Calculate the amount of bottom reinforcement required in the permanent case. Use EC2 safety factors.

Answers:

Quiz 1

(a)

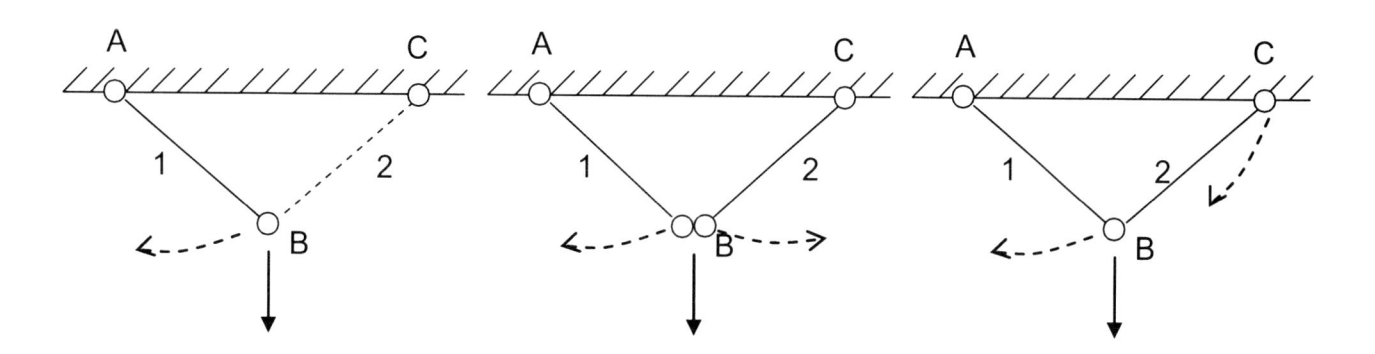

| Bar 2 removed: bar 1 swings to vertical position | Constraint between the bars removed: both bars swing to vertical position | Support C removed: bar 1 swings to the vertical position followed by bar 2 |

For (b) consider the removal of any member, say EB. The structure becomes unstable.

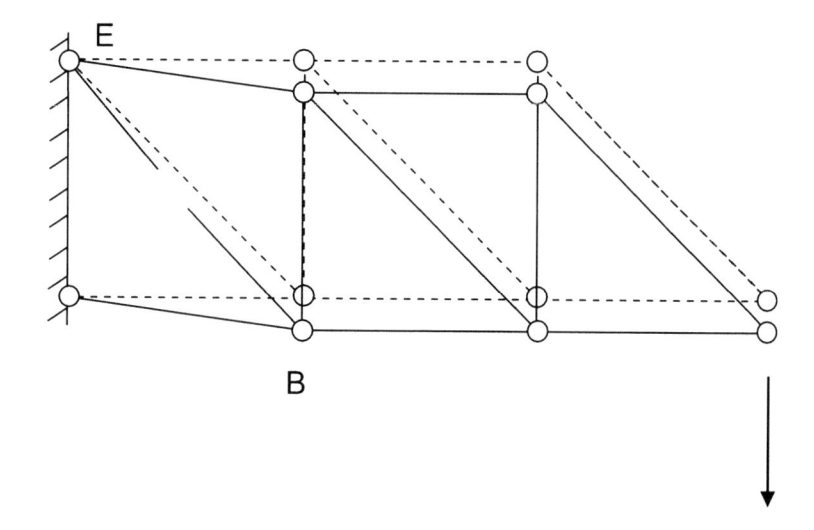

Quiz 2:

- Both structures are statically determinate.
- Structure 1(a) should definitely be avoided if possible as the consequences of any failure may be severe.
- Structure 1(b) may be used as the consequences are not likely to be as severe.

Quiz 3:

- The structure is statically determinate. It is best avoided.
- If it must be used, e.g., for architectural necessity, temperature differential or support settlement reasons, then there must be adequate load sharing ability if a truss fails.
- An out-of-plane connection is essential.

Quiz 4:

Approximate analysis can proceed by assuming positions for the points of contraflexure of the bending moment diagram, inserting pins in these positions. Sufficient pins are inserted into the beam or frame to make it statically determinate.

To make the beam in this question statically determinate assume points of contraflexure are $0.8L$ from *A* and *C*.

```
w
↓↓↓↓↓↓↓↓↓↓↓↓↓↓↓↓↓↓↓↓↓↓↓
━━━━━━━━━━━━━━━━━━━━━━━
↑           ↑        ↑
A   0.8L    B  0.2L  C
```

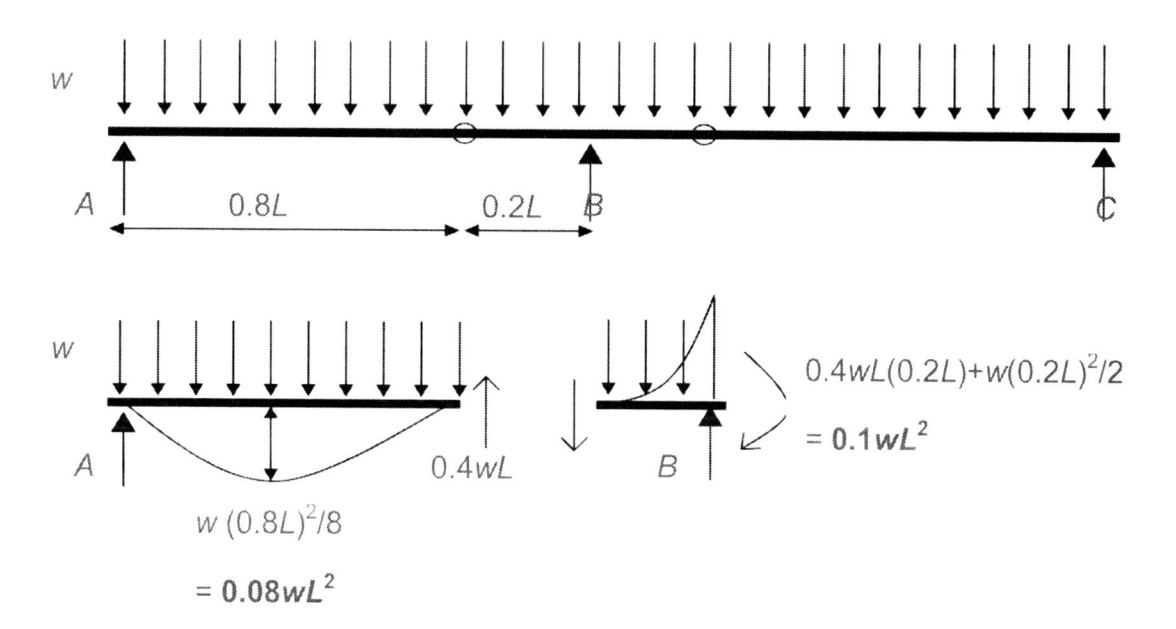

```
w
↓↓↓↓↓↓↓↓↓↓↓                    ↓↓↓
━━━━━━━━━━━                    ━━━
↑                ↑              ↑
A              0.4wL            B
   w(0.8L)²/8                 0.4wL(0.2L)+w(0.2L)²/2
   = 0.08wL²                    = 0.1wL²
```

(Remember that to ensure safety, usually all that is needed is a check on equilibrium!)

Quiz 5:

The probability of something happening is just a statement of its likely frequency. What we should be asking is "what is the risk?" Only by knowing the risk can we deal with the possibility of the event. As stated in the introduction,

Risk = Probability x Consequences.

Thus although the **probability** of a structural collapse is low (perhaps $1/10^7$ per year), the **risk** is still high.

To answer this question we should multiply the probability by the consequences.

A: 999/1000 x $1 = $0.999

B: 1/1000 x $10,000 = -$10

Total of A + B = -$9.001

Overall result is negative. Therefore, do not trade.

Quiz 6:

We are required only to estimate the natural frequency. For this we will use the approximation that natural frequency f is $18/(\delta)^{0.5}$. The applied load is $mg = 31 \times 9.81 = 304$ N/m = 0.3 N/mm

This load is applied to the cantilever as a static UDL on the whole span.

Maximum deflection at the end is $\delta = wL^4/8EI = 0.3 \times 10,000^4 / 8 \times 205,000 \times 44,390,000 = 41$ mm.

$f = 18/(41)^{0.5} = 2.8$ Hertz

(In a real structure the natural frequency of a long-span floor beam is calculated under its dead load and some proportion of live load, and compared to 4-6 Hz. If comfortably above that figure then human-induced resonance is unlikely. A natural frequency of at least 1 Hz is probably necessary for tall buildings (as gusts of this frequency are likely). Long span canopy roofs or bridges need a wind-tunnel examination to assess possibility of wind-induced resonance.)

Quiz 7:

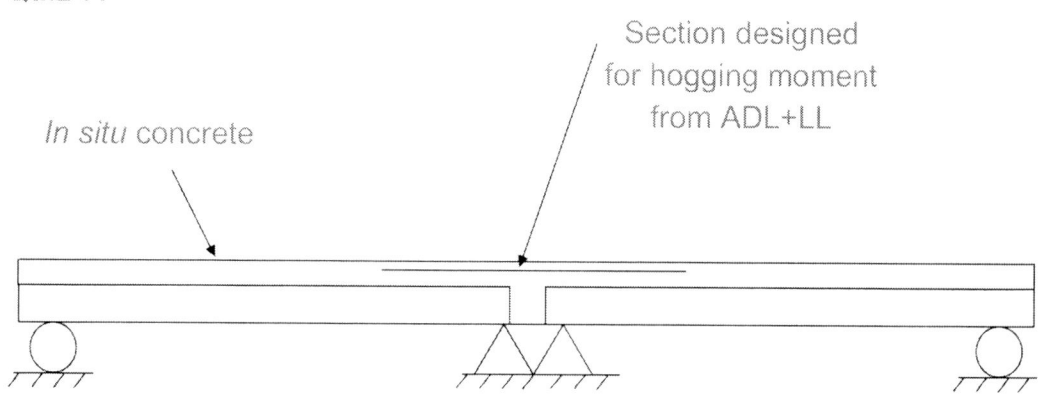

In situ concrete

Section designed for hogging moment from ADL+LL

Thermal Expansion/Contraction: Catered for using roller supports at A and C.

Advantages: Redundancy; no movement joint at B.

Disadvantages: Support movement at A or C induces moments etc, thus the beam should be as slender as possible.

Quiz 8:

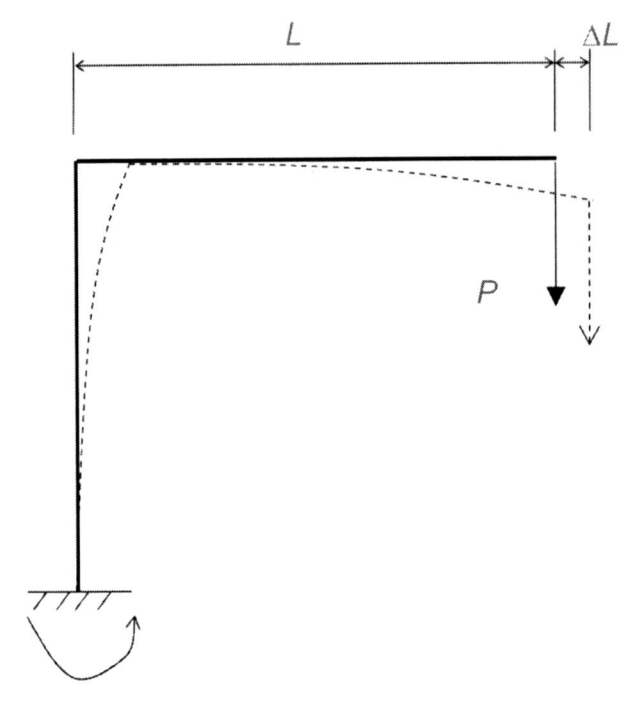

M_1 = First-Order Moment = PL

$M_{TOT} = P(L + \Delta L) = M_1 (MF)$

Expression for M_{TOT} an approximation as frame will deflect further as a result of increased moment M_{TOT}

Quiz 9:

The first order foundation moment M_1 is obtained by Ignoring deflection: M_1 = 0.8x9x16x4.5 = <u>518.4</u> kNm

Assuming a linear deflected shape => if the roof moves H/500 = 9,000/500 = 18 mm, then level 3 moves 12 mm, and level 2 (first suspended level) moves 6 mm.

Thus additional Moment due to deflection= (0.25x24)x32x16x(0.018+0.012+0.006) = 110.6 kNm

Hence total moment the foundation must resist is 629.0 kNm.

Thus modification factor 629/518.4 = 1.21

This MF is excessive. The stiffness of the lateral load resisting system should be increased so that MF is not greater than 1.1. It can be shown that this means that in this case the deflection should be not greater than about 11mm at the top.

Quiz 10:

Close inspection of the supports showed that the right hand support had partially rotted away and so the column was now resting on the two remaining supports. Thus half of the column was acting as a cantilever.

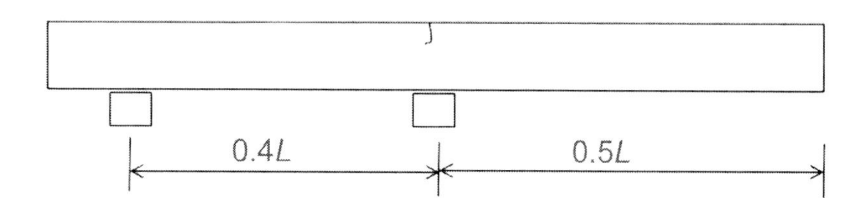

It can be shown that the maximum bending moment before the modification was $0.075wL^2$ and after the modification and the settlement $0.125wL^2$ where w is the uniformly distributed self-weight. Thus there was an approximately 66% increase in moment on the section.

Clearly by storing the column in a statically indeterminate way can be unfavourable if support settlement is a risk. A ductile material can cope with small settlements but not a brittle one.

Quiz 11:

The truss is statically determinate. Thus an alternative means of support must be found in the event the main load-paths are lost (for whatever reason). Introduce purlins to support the roof deck; detail the purlins as continuous over all the trusses so that in the event of an accident the trusses can hang as a catenary. Thus continuous ties perpendicular to the plane of the truss is all that is necessary. In this way the immediate neighbours of a truss can be designed so that they can support the failing truss. So the engineer must design the typical truss for the extra load that this implies.

The lightweight roof is subject to uplift from the wind. Thus the tie of the roof can not be allowed to go into compression. Thus it must be pre-tensioned, e.g., by pre-compressing the king-post.

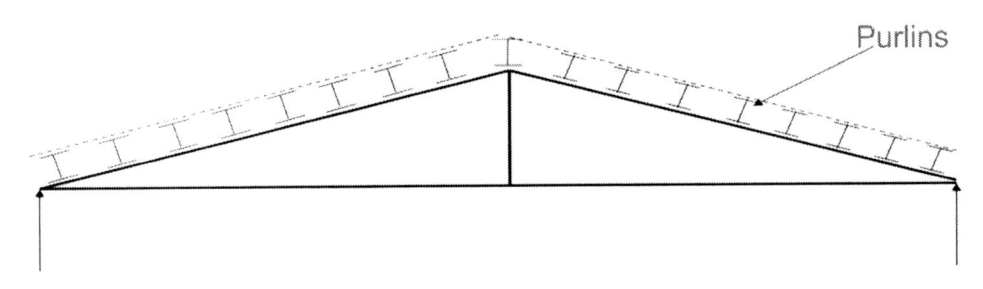

An additional measure that could be used to improve the redundancy further is to use a pair of steel round ties as the truss tie. This would also allow the use of smaller pieces of steel and so make connections easier to detail.

Quiz 12:

Ignoring the fact that the supports are fixed, means that the support moment is being ignored. This is the same as saying that the beam is to be designed based on a redistribution of 100%. If design is based on a mid-span moment of $wl^2/8$ then the **equilibrium** requirement has been satisfied and so the beam is actually **safe** under this load.

But this is true only if there is adequate ductility. The concrete must not crush before the moment is redistributed. (This is why there is a limit on the value of *x/d* in EC2). The beam may be safe but it is *not* **serviceable**: no top steel would be provided over the supports and so wide cracks would open up there. Clearly EC2 would not allow such a beam as, at most, redistribution can be 30% according to the code. (This of course is conservative. It is well established (106) that full redistribution of moments in under reinforced beams with comparatively light reinforcement is possible.)

Quiz 13:

Robertson's formula:

Modification Factor $(MF) = 1/(1-WR/Q)$

Building weight $W = 190 \times 9.81 \times 31 \times 90 \times 234 = 1217 \times 10^6$ N.

Design drift $R = 1/400$.

Wind pressure is 1,400 N/m^2 on <u>narrow</u> face of building (31 m).

Thus $Q = 1,400 \times 31 \times 234 = 10.1 \times 10^6$ N.

Hence MF = = $1/(1-WR/Q) = 1/(1 - 1,217 \times 10^6/(400 \times 10.1 \times 10^6)) = 1.43$

Clearly this is excessive (limit about 1.1, i.e., second order effects around 10%)

Quiz 14:

Using results of research in reference 45 (see Bailey's Crossroads case study):

Ultimate capacity (28-day) according to EC2 = 1.35(DL+ADL) + 1.5LL = 1.35×(5.0+1.5) + 1.5×2.5 = 12.5 kN/m^2

Demand = 2.25DL = 2.25x5 = 11.25 kN/m² < 12.5 kN/m²

Hence scheme is probably okay provided the slab concrete can reach the specified 28-day strength in 14 days (2 cycles).

Above calculation ignores construction live load and weight of formwork.

Quiz 15:

Estimated design capacity = 1.35DL + 1.5LL

225 mm thick slab => DL = 5.6 kN/m².

Assume LL used for design = 2.5 kN/m².

Assume level 3 slab was 225 mm thick and carried no LL at time of collapse of level 4.

Thus 0.35(5.6) + 1.5(2.5) = 5.7 kN/m² over static DL is available to resist collapse.

Thus according to EC2 level 3 has sufficient resistance if the dynamic multiplier is no more than about 2.

Quiz 16:

Measures:

- Upper structure:
    - Ensure all ties placed as required by code.
    - Ensure cover as required by code.
- Level 2 transfer structure:
    - There are two structural options:

(1) Ensure the central column at level 1 can resist the effect of likely accidental transverse load (34 kPa) from any direction: note bracing members should be designed for this load too (upward critical).

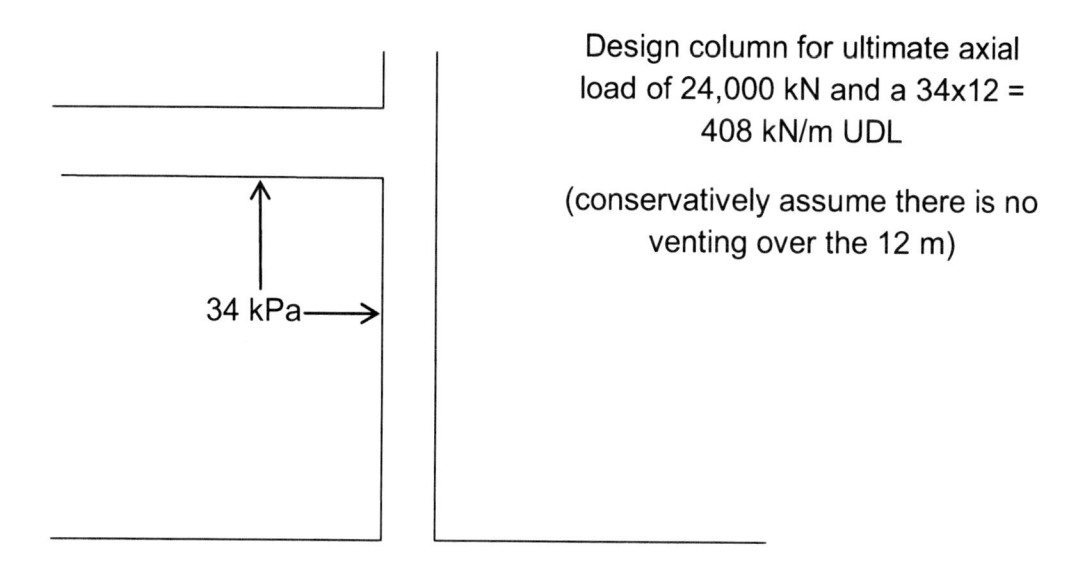

Design column for ultimate axial load of 24,000 kN and a 34×12 = 408 kN/m UDL

(conservatively assume there is no venting over the 12 m)

34 kPa →

Alternatively,

(2) Design building so that the central column at level 1 can be removed. Consider a static removal of column (i.e. ignore dynamic effects). Assume the transfer beam acts as catenary (continuous bottom reinforcement); Detail columns to allow them to act as hangers (thus linking levels). Design other columns to resist extra axial force.

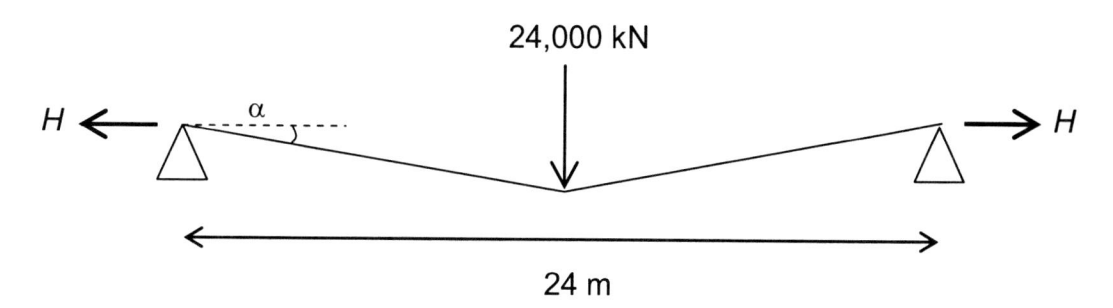

24,000 kN

H ←   α   → H

24 m

Experiments have shown that the ultimate deflection of the member can be taken as about 0.1L

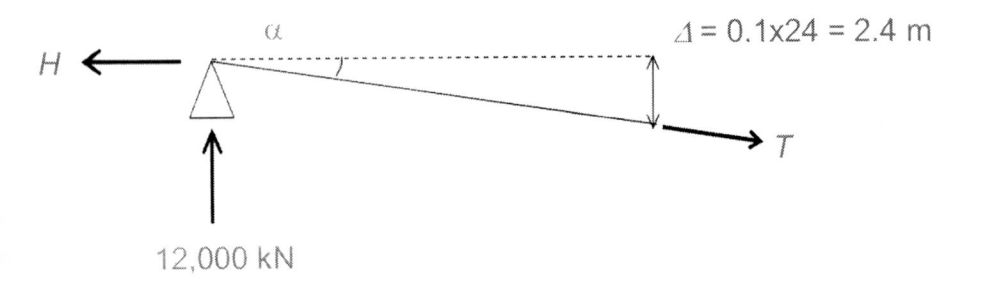

$\Delta = 0.1 \times 24 = 2.4$ m

12,000 kN

$T \sin \alpha = 12{,}000$; $T \cos \alpha = H$

$\Rightarrow H = 12{,}000/\tan \alpha = 12{,}000/0.2 = 60{,}000$ kN;

Assume that all 17 floors can have beams which act as catenarys

$\Rightarrow A_s = 60{,}000{,}000/(460 \times 17) = 7{,}673$ mm^2; Say 10T32s per level (8,040 mm^2).

It should be noted that this is not extra reinforcement. Our assumption is that the concrete has already failed. Thus the reinforcement for moment that is provided is available for use as catenary reinforcement.

Other (non-structural) measures:

- Use non-laminated glass so that ventilation reduces effects of internal explosions.
- Use cylinders rather than piped gas (less gas can escape before ignition).

"For structures of high significance and exposure, direct design methods are generally preferable to prescriptive design rules"

(Uwe Starossek author of "Progressive Collapse of Structures", 2009)

Quiz 17:

Mean concrete strength, f_{cm} = 38 N/mm^2 (Table 3.1)

Assume mean steel strength = 10% above characteristic yield.

Axial capacity without safety factors = 0.85x38x300^2 + 4x491x1.1x500 = 2907 + 1080 = 3987 kN

Axial capacity including material safety factors = 0.57x30x300^2 + 4x491x0.87x500 = 1530 + 854 = 2384 kN

Safety Factor considering materials alone = 3987/2384 = 1.67

Safety Factor considering loading alone = (1.35+1.5)/2 = 1.43

Overall Safety Factor = 1.67x1.43 = 2.4

Quiz 18:

Shear using characteristic strengths:

Check: span/depth = 2000/400 = 5 > 3 not a 'deep' beam => okay

Ultimate UDL w = 1.35x2.25+1.5x200 = 303 kN/m so V_{max} = Reaction = 303 kN

Check shear stress at supports:

$v_{max,z}$ = $V_{max}/b_w 0.9d$ = 303,000/200x0.9x400 = 4.21 N/mm^2,

$v_{Rd,max\ cot\ \theta = 2.5}$ = 4.63 N/mm^2 > 4.21 N/mm^2.

Thus beam can be reinforced to resist these stresses. Assume cot θ = 2.5

Maximum shear d from support = V_{Ed} = 182 kN

$v_{Ed,z} = V_{Ed}/(b_w z) \approx V_{Ed}/(b_w 0.9d) = 182 \times 10^3/(200 \times 0.9 \times 400) = 2.53$ N/mm^2

Provide reinforcement such that $A_{sw}/s \geq v_{Ed,z} b_w/(f_{ywd} \cot \theta)$ ...(1)

Using $\phi 8$ links (2 legs) => $A_{sw} = 100$ mm^2

=> $s = A_{sw} f_{ywd} \cot \theta / v_{Ed,z} b_w = 100 \times 0.87 \times 500 \times 2.5/2.53 \times 200 = 215$ mm

Notice that the strength of concrete struts do not govern so estimate does not depend on concrete strength (see...(1))

Shear using mean material strengths:

As above except $f_{ywd} = 550$ N/mm^2

Thus factor of safety for shear = $(550/500)(1.15)(1.5) =$ **1.9**

Flexure using characteristic strengths:

Tensile force, $T = 0.87 \times 500 \times 2 \times 491 = 427.1$ kN

Depth of compression stress block, $s = 427.1 \times 10^3/(0.57 \times 40 \times 200) = 94$ mm

Lever arm, $z = d - s/2 = 400 - 94/2 = 354$ mm

Flexural moment of resistance = $Tz = 427.1 \times 10^3 \times 354 = 151.2$ kNm

Flexure using mean strengths:

Tensile force, $T = 550 \times 2 \times 491 = 540.1$ kN

Depth of compression stress block, $s = 540.1 \times 10^3/(0.85 \times 48 \times 200) = 66$ mm

Lever arm, $z = d - s/2 = 400 - 66/2 = 367$ mm

Flexural moment of resistance = $Tz = 540.1 \times 10^3 \times 367 = 198.2$ kNm

Factor of Safety for flexure = (198.2/151.2)(1.5) = **1.97**.

Comment: Having similar factors of safety for flexure and shear implies that the code (EC2) is assuming any such failure will be non-catastrophic. This is likely to be true only in continuous beams. Any statically determinate structure (or part of a statically indeterminate one, e.g. a cantilever overhang) needs a substantially larger factor of safety against shear failure than flexural.

Suggested procedure for shear design:

Upper limit to M_{flex} = 198.2 kNm

Corresponding $V = V_{design} = wL/2$ where $M_{flex} = wL^2/8$ => $w = 8M_{flex}/L^2 = 8 \times 198.2/2^2 = 396.4$ kN/m

i.e., $V_{design} = 396.4 \times 2/2 = 396.4$ kN

Additional Factor of Safety for shear = 396.4/303 = 1.3

Quiz 19:

Mid-span deflection under working load of DL + LL = 10.5 kN/m:

$$\frac{5}{384}\frac{wl^4}{EI} = \frac{5}{384}\frac{10.5 \times 15{,}000^4}{210 \times 10^3 \times 273 \times 10^6}$$

= 121 mm = $L/124$ (Deflection under LL alone $L/289$)

Now, assume that this 'bowl' can fill with water:

Weight of water = average depth is about 60 mm = 0.6 kN/m^2

This is less than the 0.75 kN/m^2 assumed LL so ponding is not likely to be a problem.

↕ 121 mm

Quiz 20:

Check using a masonry code, e.g. EC6 (BS 2005c) or similar. Single skin => t = 100 mm.
Slenderness ratio SR = 0.75x3000/0.1 = 9/40, e = 0.05t, ϕ_i = 1-2x0.05 = 0.9

f_b = 50 N/mm^2 and f_m = 12 N/mm^2;

General purpose mortar => γ_m = 2, α = 0.7, and β = 0.3

$f_d = 0.4 f_b^\alpha f_m^\beta / \gamma_m$ = 6.4 N/mm^2

$N_{Rd}/t = \phi_i f_d$ = 0.9x6.4 = 5.8 N/mm^2

Thus the allowable working stress (using γ_f of say 1.5) is 5.8/(1.5) = 3.9 N/mm^2

Thus 3 N/mm^2 is a significant part of the expected capacity perhaps likely to be too much.

Quiz 21:

Accident Load = (DL + 0.5LL) = (0.2x25 + 1 + 0.5x2.5) = 7.25 kN/m^2

Capacity of T13 = $A_s f_y$ = 133x500 = 66.5 kN

Thus no. of T13s = 7.25x6x6/66.5 = 3.9

Provide a minimum of 4T13 bars.

A plan of the reinforcement is shown below.

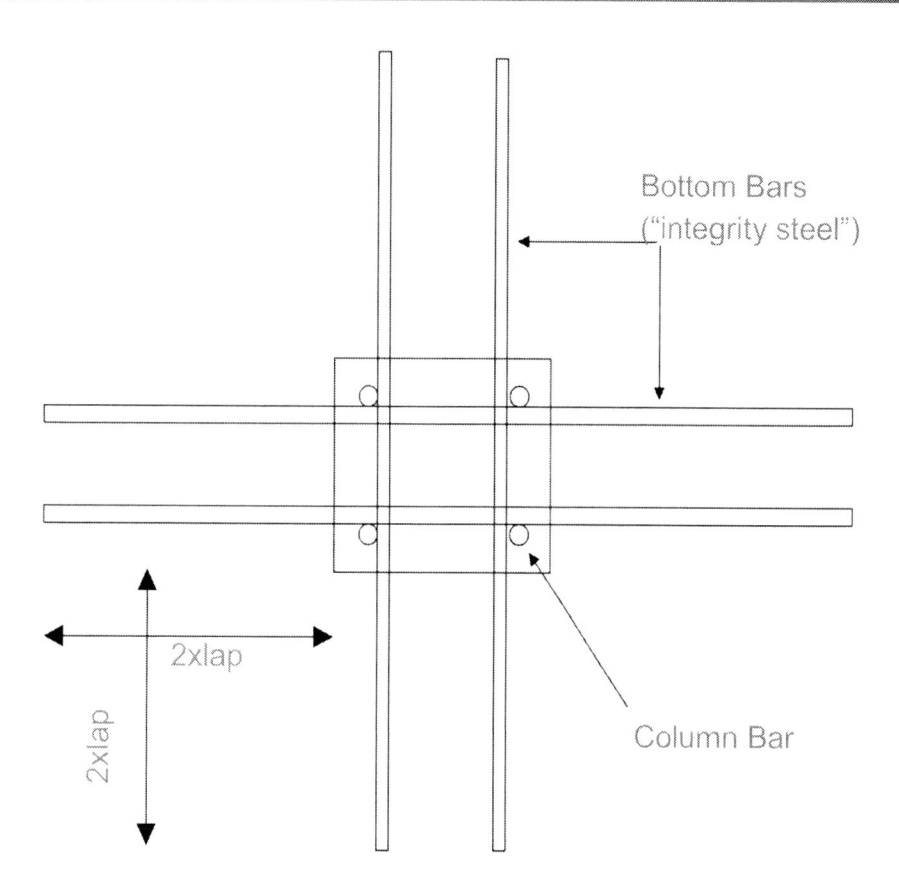

References

ACI-318-08, *Building Code Requirements for Structural Concrete*, American Concrete Institute, 2008.

Alexander, S., *Axial shortening of concrete columns and walls*, Concrete, March 2001, pp 36-38.

Angelo, W., *Collapse Report Stirs Debate on Epoxies*, Engineering News Record, 18 July 2007.

Ashby, M. & Jones, D., *Engineering Materials 2*, 2nd edition, Butterworth-Heinemann, 1998.

Bate, S., *Report on the failure of roof beams at Sir John Cass's Foundation and Red Coat Church of England Secondary School, Stepney*, Building Research Station, BRE, 1974.

Bazerman, M. & Watkins, M., *Predictable surprises: the disasters we should have seen coming, and how to prevent them*, Harvard Business, 2008.

Berthier, *The collapse of terminal 2, Charles de Gaulle Airport*, Technical presentation of commission (available at www.equipement.gouvfr).

Billington, D. P., *Power, Speed, and Form*, Princeton University Press, 2006.

Billington, D. P., *The tower and the bridge: the new art of structural engineering*, Princeton University Press, 1983.

Blackhall, S., *World's Greatest Blunders*, Octopus Books, London, 1989.

Blockley, D. I., *The Nature of Structural Design and Safety*, Chichester, Ellis-Horwood, 1980.

BS EN 1992-1-1:2004, *Eurocode 2: Design of concrete structures* — Part 1-1: General rules and rules for buildings, British Standards Institution. 2005.

BS EN 1993-1-1:2005, *Eurocode 3: Design of steel structures* — Part 1-1: General rules and rules for buildings, British Standards Institution. 2005.

BS EN 1996-1-1:2005, *Eurocode 6: Design of masonry structures*, British Standards Institution. 2005.

BS5950: Part 1: 2000, Structural Use of Steelwork in Building, British Standards Institution, London, 2000.

BS8110:1997, The Structural Use of Concrete-Part 1: Code of Practice for Design and Construction, British Standards Institution, London, 1997.

Burgoyne, C., *Why did Palau Bridge Collapse?*, The Structural Engineer, June 2006.

Cambell, R,. *Builder Faced Bigger Crisis Than Falling Windows*, Boston Globe, 3 March 1995.

Campbell, P., *Learning from Construction Failures: Applied Forensic Engineering*, Whittles Publishing, London, 2001.

Chapman, J., *Collapse of the Ramsgate Walkway*, The Structural Engineer, January 1998.

Chew, M. et al. *Building Facades: a guide to common defects in tropical climates*, World Scientific, 1998.

Chiles, J., *Inviting Disaster: Lessons from the edge of technology*, Harper Business, USA, 2002.

Chilton, J., *The Engineer's Contribution to Contemporary Architecture: Heinz Isler*, Thomas Telford, 2000.

Colaco, J., Ford, W., Robertson, G., *Complete retrofit of a 47-story steel building for wind loads*, CTBUH Review, Volume 1, May 2000, pp 30-37.

Collins, M.P. et al, *The Failure of an Offshore Platform*, Concrete International, August 1997.

Commission, *Inquiry into the collapse of a portion of the de la Concorde overpass*, October 15, 2007.

Dallard, P. et al., *The Millennium Bridge, London: Problems and Solutions*, The Structural Engineer, Vol 79, No. 8, April 2001.

Delatte, N., *Beyond Failure: Forensic Case Studies for Civil Engineers*, ASCE Press, 2009.

Discussion to Hulme et al., The Structural Engineer, Vol 72, No. 15, August 1994.

References

Discussion to Woodward and Williams, Proceedings of the Institution of Civil Engineers, Part 1, No 86, Dec 1989.

Dowling, P. et al., *Structural Steel Design*, Butterworths, 1988.

Duffey, R. & Saull, J., *Know the Risk*, Butterworth Heinemann, London, 2003.

Dutt, A. & George, T., *Case history of wind-induced failure of a bridge structure*, International Conference on Case Histories in Structural Failures, March 1989, Singapore.

Elliot, K. S., *Multi-storey precast concrete framed structures*, Blackwell Science, 1995.

Elstner, R. & Hognestad, E., *Laboratory Investigation of Rigid Frame Failure,* Journal of American Concrete Institute, Vol. 28, No. 7, Jan 1957.

ETABS, Integrated Building Analysis and Design Software, Computers and Structures Inc.

Feld, J. & Carper, K.L., *Construction Failures*, Second Edition, John Wiley & Son, 1997.

Ferguson, E., *Engineering and the Mind's Eye,* MIT Press, 1992.

Fleddermann, C., *Engineering Ethics*, 2nd edition, Pearson, 2004.

Fontana, M., *Corrosion Engineering*, 3rd edition, Mc-Graw-Hill, 1986.

Francis, A. J., *Introducing Structures*, John Wiley, 1989.

Gardner, N. J. et al, *Lessons from the Sampoong department store collapse*, Cement and Concrete Composites, Vol. 24, Issue 6, Dec 2002.

Gerstein, M., *Flirting with Disaster: why accidents are rarely accidental*, Union Square Press, New York, USA, 2008.

Gillespie, A., *Twin Towers: The Life of New York City's World Trade Centre*, Penguin, USA, 2002.

Glanz, J., *Wounded Buildings Offer Survival Lessons*, New York Times, December 4, 2001.

Gordon, J. E., *The New Science of Strong Materials*, 2nd Edition, Princeton Science Library, 1988.

Great Britain, Office of the Deputy Prime Minster, *The Building Regulations 2000*: Approved document A-Structure, 2004 ed, London: The Stationery Office, 2004.

Grossman, J., *Slender Concrete Structures-the new edge*, ACI Structural Journal, Vol. 87, No. 1, January-February 1990, pp. 39-52.

Hampton, T., et al, *Multiple reports, video could shave months off investigation of fatal Minnesota bridge collapse*, Engineering News Record, McGraw-Hill, 2 Aug 2007.

Hawkins, N. & Mitchell, D. *Progressive Collapse of Flat Plate Structures*, ACI Journal, Vol. 76, pp775-808, July 1979.

Heyman, J., *Basic Structural Theory*, Cambridge University Press, Cambridge, UK, 2008.

Heyman, J., *Elements of stress analysis*, Cambridge University Press, Cambridge, UK, 1982.

Heyman, J., *The Science of Structural Engineering*, Imperial College Press, 1999.

Hulme, T. et al., *The collapse of the Hotel New World, Singapore: a technical inquiry*, The Structural Engineer, Vol. 71, No. 6, March 1993.

Hunter, *NSTB finds fractured gusset plates in I-35W span*, Engineering News Record, 16 January 2008.

Hurd, M.K., & Courtois, P.D., *Method of analysis for shoring and reshoring in multi-storey buildings*. Proceedings of 2nd international conference on forming economic buildings, USA. 1986. Paper SP-90-8.

Ichniowski, T., *NSTB blames Minnesota Bridge collapse on gusset plate design error*, Engineering News Record, McGraw-Hill, 13 Nov 2008.

James F. Lincon Arc Welding Foundation, *The Procedure Handbook of Arc Welding*, 14th Ed., 2000.

Jennings, A., *Structures: from theory to practice*, Spon Press, London, 2004.

Kaminetzky, D., *Design and Construction Failures,* McGraw-Hill, 1991.

Kletz, T., *Learning from accidents*, 3rd edition, Gulf Professional Publishing, 2001.

References

Kotsovos, M.D. & Pavlovic, M.N., *Ultimate Limit-State Design of Concrete Structures*, Thomas Telford, 1999.

LePatner, B, et al, *Structural and Foundation Failures*, McGraw-Hill, 1982.

Levy, M. & Salvadori, M., *Why Buildings Fall Down,* Norton, 1992.

Lichtenstein, A., *The Silver Bridge Collapse Recounted*, Journal of Performance of Constructed Facilities, ASCE, November 1993.

MacLeod, I., *Modern Structural Analysis: Modelling Process and Guidance*, Thomas Telford, 2005.

Magnus, R., *Report of the committee of enquiry into the incident at the MRT circle line worksite that led to the collapse of Nicholl Highway on 20th April 2004*. Two volumes. May 2005.

Mark, R., *Light, wind, and structure*, The MIT Press, 1990.

Martin, M. & Schinzinger, R., *Ethics in engineering*, 3rd edition, Mc-Graw-Hill, 1997.

Mast, R., *Roof Collapse at Antioch High School*, Journal of the Prestressed Concrete Institute, 1980.

McDonald et al, *A spectacular collapse: the Koror-Babeldoub (Palau) balanced cantilever post-tensioned bridge*, Proceedings 27th Our World in Concrete and Structures, XXI:57-68, Aug 2002.

McEvily, A. J., *Metal Failures: mechanisms, analysis, prevention*, John Wiley & Sons, New York, 2002.

McKaig, T., *Building Failures: Case studies in construction and design*, McGraw-Hill, New York, 1962.

Miga. A., *Epoxy creep factor in Big Dig death*, Engineering News Record, 10 July 2007.

Mlakar, P., et al, *The Pentagon Building Performance Report*, ASCE/SCI, USA, 2003.

Moncarz, P. et al, *Engineering Process Failure-Hyatt Walkway Collapse*, Journal of Performance of Constructed Facilities, May 2000.

Morgenstern, J., *The Fifty-Nine-Story Crisis*, Journal of Professional Issues in Engineering Education and Practice, ASCE, 1997.

Murta-Smith, E., *Space Structural Integrity*, Space Structures 4, Proceedings of the Fourth International Conference on Space Structures, 1993.

Newman, A., *Structural Renovation of Buildings: methods, details and design examples.* McGraw-Hill, New York, USA, 2001.

NIST, *WTC 7 Technical Briefing*, National Institute of Standards and Technology, draft of final report available from NIST website (http://wtc.nist.gov), 2008.

Perrow, C., *Normal Accidents*, Princeton University Press, 1999.

Perrow, C., *The Next Catastrophe*, Princeton University Press, USA, 2007.

Petroski, H., *Design Paradigms*, Cambridge Press, 1994.

Petroski, H., *Engineers of dreams: Great bridge builders and the spanning of America*, Vintage Books, 1995.

Petroski, H., *Invention by design*, Harvard University Press, 1996.

Petroski, H., *Past and Future Failures*, American Scientist, Vol. 92, November-December 2004.

Petroski, H., *Remaking the world: adventures in engineering*, Vintage Books, 1997.

Petroski, H., *Success through failure: the paradox of design*, Princeton University Press, USA, 2006.

Petroski, H., *The Minneapolis Bridge*, American Scientist, Vol. 97, No. 6, November-December 2009.

Petroski, H., *To Engineer is Human,* Vintage, 1992.

Post, N., *Engineers slam failed joint detail*, Engineering News Record, McGraw-Hill, 3 May 2007.

Ransom, W., *Building Failures,* E&F Spon, 1981.

Ratay, R., *Forensic Structural Engineering Handbook*, McGraw-Hill, New York, USA, 2010.

References

Reason, J., *Managing the risk of organizational accidents*, Ashgate, UK, 1997.

Regan, P.E, *Behaviour of reinforced concrete flat slabs*, Report 89, CIRIA, 1981.

Reina, P., *Failed airport structure's fate to be decided by mid-April*, Engineering News Record, 28 February 2005.

Robertson, L., et al., *Tall Buildings Criteria and Loading*, Vol CL (Monograph on Tall Buildings, Council for Tall Buildings and Urban Habitat), ASCE, 1980.

Robertson, L., See, S., *Preliminary Design of High-Rise Buildings*, Building Structural Design Handbook, Ed. White & Salmon, Wiley, NY, 1987.

Ross, S., *Construction Disasters,* McGraw-Hill, 1984.

Salmon, M., Fixing Failures, Construction Fixings Association Articles, CFA, (www.fixingscfa.co.uk).

Schlager, N., *When Technology Fails: Significant Technological Disasters, Accidents, and Failures of the Twentieth Century*, Gale Research, Detroit, 1994.

Shepherd, R. & Frost, J., *Failures in Civil Engineering-Structural, Foundation and Geoenvironmental Case Studies*, ASCE, 1995.

Silby, P., & Walker A., "Structural Accidents and their causes", Proc. ICE, Vol. 62, pp. 191-208, 1977.

Simu, E. & Scanlan, R., *Wind Effects on Structures*, 3rd edition, Wiley, 1996.

Starossek, U., *Progressive Collapse of Structures*, Thomas Telford, UK, 2009.

Taranath, B., *Steel, concrete and composite design of tall buildings*, 2nd edition, McGraw-Hill, 1998.

Teh, H.S., *An Overview Of The Cast-in-place Balanced Cantilever Method Of Bridge Construction*, Journal of Institution of Engineers, Singapore, Vol.29, No.1, April 1989.

The Institution of Structural Engineers, *Safety in tall buildings and other buildings with large occupancy,* The Institution of Structural Engineers, London, July 2002.

Thean, L., et al, *Report of the Inquiry into the collapse of Hotel New World*, 1987.

Thung, S., K., *Doomed From the Start* (Findings on the collapse of multi-purpose hall in Compassvale Primary School), Building Construction Authority, Singapore, 2000.

Tomasetti, R. & Abruzzo, J., Protective Design of Structures, Chapter 22 of *Building Security* by B. Nadel, McGraw-Hill, 2004.

Trahair, N. et al. *The behaviour and Design of Steel Structures to EC3*, 4th Ed., Taylor & Francis, London, 2008.

Various, *Forensic engineering: a professional approach,* Proceedings of international conference, ICE, London, 1998.

Wearne, P., *Collapse: When Buildings Fall Down*, TV Books, USA, 2000.

WJE, *David L Lawrence convention centre-investigation of the 5 February 2007 collapse*, Pittsburgh, Wiss, Janney, Elstner Associates Inc, 2008.

Wood, J.G.M., *Implications of the collapse of the de la Concorde overpass*, The Structural Engineer, 8 January 2008.

Wood, J.G.M., *Paris airport terminal collapse*, The Structural Engineer, 1 March 2005.

Wood, J.G.M., *Piper's Row car park collapse: Identifying risk*, Concrete, October 2003.

Wood, J.G.M., *Pipers Row Car Park, Wolverhampton: Quantitative study of the causes of the partial collapse on 20th March 1997*, Structural Studies & Design Contract Report to HSE (www.hse.gov.uk).

Woodward, R. & Williams, F., *Collapse of Ynys-y-Gwas bridge, West Glamorgan*, Proceedings of the Institution of Civil Engineers, Part 1, No 84, Aug 1988.

Zallen Engineering, *Deficiencies in bracing of lift slab buildings,* Zallen Engineering Issue No. 11, March 2004. Online edition available at http://www.zallenengineering.com/On-Line_Issues/OL-11.pdf

Appendix 1:

Hambly's Paradox (Heyman 1999)

Consider a three-legged stool, shown in Figure A1 below. There is a centrally applied point load, W.

If it is assumed the legs can only take compression then the force in each leg F_1 can be determined using statics. Thus a three-legged stool is statically determinate. Vertical equilibrium and symmetry mean each leg takes equal force. Taking moments about say A means that F_1 is known to be $W/3$.

Now consider the four-legged stool shown in Figure A2. Clearly it is statically indeterminate. Call the force in each leg F_2. A conventional elastic analysis (i.e., taking legs as equally supported) gives $F_2 = W/4$. But to get this figure we have perhaps unconsciously made additional assumptions. We will first demonstrate the paradox: suppose the material is prone to brittle failure (either the material itself is brittle or it is ductile, as demonstrated by a tension test, but can buckle once the compression becomes critical). Consider now that one of the legs is short. If it does not touch the ground the force in it is clearly zero. Now from equilibrium the force in the leg diagonally opposite is also zero. Thus we have only two legs effectively supporting the load. So each leg must be designed for $F_2 = W/2$. This is the paradox: we have increased the number of legs but each one now must be designed for a higher force. The resolution of the paradox is in our assumption of ductility. We can carry out our elastic analysis without thinking about the boundary conditions and $W/4$ is safe only if we have ductility. Initially the legs take half the load but once they begin to yield the other two legs are able to take-up some load. In the final case we have an even spread onto the four legs. In order to have ductility we must have a material that behaves in a ductile way and there should be no possibility of buckling. Thus the Factor of Safety (FS) against buckling needs to be higher than that against yield. The minimum FS against buckling here is based on our elastic analysis which gives an $F_2 = W/4$. Thus we need a FS of at least 2 to avoid premature failure of the legs.

This illustrates the unstated assumption whenever we use elastic theory (which we invariably do) that the material is ductile and there is no chance of buckling or other kind of brittle failure (this is achieved by making the FS high).

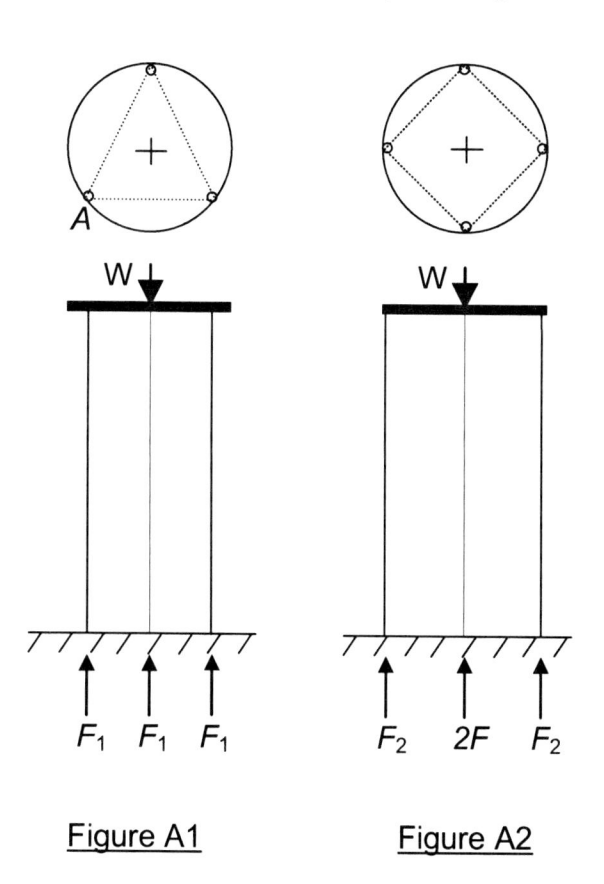

Thus designing based on the elastic solution is okay if the material and structural behavior is ductile as the elastic solution is in equilibrium with the applied loads.

Figure A1

Figure A2

Appendix 2

Suggested Checking Procedure: Tall buildings, (say above 10 storeys)

(a) Preliminary checks:

1. Check ETABS model (under "Analyze" select "Check Model").

2. Run ETABS. Ensure there are no errors or warnings in output. Check *.$OG file for error messages. (Warnings of calculations being in error less than 11 significant digits are likely to be okay). Note values of drift measured at the top of the building under 50-year wind.

3. Run spreadsheet "preliminary design of multi-storey buildings". Input values of drift. This will indicate items that may need further attention. The spreadsheet makes preliminary estimates of the following:

    - Modification Factor for buildings designed using a drift criterion.

    - Accelerations at the top of the building.

    - Building's susceptibility to vortex shedding.

    - Factor of safety against stability failure.

    - Amount of shear wall bracing required to limit $P\Delta$ effects.

    - Adjusted period of the building allowing for $P\Delta$ effects.

    - Likely wall stresses at the base if the core is rectangular.

    - Sensitivity of the design to base rotation.

(b) Check Structural Analysis:

1. Model Validation (i.e., check if model can represent reality):

    - Check floors are modeled as shell elements,

    - Check shear walls modeled as shell elements so bending torsion can be modeled, (check both by checking "assignments"),

    - Check I used for shear walls: gross I appropriate if no large tensions on wall, frame interaction with walls will be overestimated if gross I of floor used (check by looking at stress in wall under DL + WL).

    - Check if "line constraint" option chosen under "assignments".

    - Check 50% slab modulus used (allows for cracking).

2. Check input data check for ETABS.

3. Check that no warning file has been generated (*.WRN).

4. Overall equilibrium check (i.e., Applied load – Reactions = Residual) Value of Residual should be small, otherwise ill-conditioning likely. Totals of applied forces and reactions available in *.OUT file. Check "% out of balance error" in "Text document". Accept results if error small ($< 10^{-5}$).

5. Restraint check (i.e., look at deformations at restrained nodes).

6. Qualitative check (i.e., look at deflected shape-should be a smooth curve, nearly a straight line, and look at 1st mode shape-it should be a straight line so that accelerations are minimized; check column load accumulation: can indicate where construction sequence needs to be checked).

7. Quantitative check (i.e., create simple checking model: cantilever same height as building with I of shear walls. Attempt to get close to wind drift values in each direction from ETABS.

8. Look at the model output to verify if wind or notional load governs the design.

9. Check meshing of shear walls. Elements should have a small aspect ratio (i.e., quadrilateral elements to be as "square" as possible.). Otherwise, there should be at least 4 elements to model bending properly.

10. Check meshing of slabs. Element nodes must coincide with columns otherwise load-transfer inadequate locally. Aspect ratio of element should be no more than 4. Quadrilateral elements should have internal angles nearly 90 degrees.

11. Carry out sensitivity check on actual model to determine significant influencing parameters, e.g., check effect of reducing interaction between core and frame by changing I of floor by 50%. Drift at top of building should increase. If not something is wrong with the model.

12. Check that torsional natural frequency and bending (i.e., translation) natural frequency are **not** similar. Preferably the 2 first modes should be bending rather than torsion.

13. Carry out sensitivity check on value assumed for modulus E as creep will reduce it.

(c) Check Structural Design:

Spot checks to ensure design implemented correctly (e.g., buckling length correctly used in steel design). Check at least minimum reinforcement used in walls and columns. Hand check on approximate reinforcement required in a shear wall and/or column. Ensure lateral shear reaction can be supported by base; basement walls can be used for this, otherwise use raking piles.

(d) Check Progressive Collapse Resistance:

e.g., ensure sufficient ties and bottom reinforcement through column cage in flat slabs; ensure continuous bottom reinforcement to allow catenary action; ensure column reinforcement can act as a tie in event of accident.

Index of cases:

Abbeystead, 162
Alexander Kielland, 122
Alfred Murrah Building, 165
Amoco Tower, 169
Antioch School, 92
Bailey's Crossroads, 139
"Big-dig" tunnel, 192
Brooklyn College, 119
Citicorp Centre, 75
Cocoa Beach Condominium, 146
Compassvale School, 51
D. L. Lawrence Convention Centre, 222
De la Concorde Overpass, 103
Ferrybridge Cooling Towers, 65
Hartford Civic Centre, 37
HKHA Blocks, 187
Hotel New World, 25
I-35W Bridge, 219
John Hancock Tower, 69
K-B Bridge, 212
Kemper Arena, 202
King's Street Bridge, 115
L'Ambiance Plaza Condominium, 148
Mianus River Bridge, 87
Millennium Bridge, 216

MRT Circle Line, 151
Paris Airport, 33
Piper's Row Car Park, 100
Point Pleasant Bridge, 84
Quebec Bridge, 129
Ramsgate Walkway, 47
Ronan Point, 157
Sampoong Department Store, 28
Schoharie Bridge, 97
Singapore Condominium, 188
Sleipner A, 42
SMRF during Northridge earthquake, 125
Station Square Shopping Centre, 209
Stepney School, 199
Sunshine Skyway Bridge, 160
Tacoma Narrows, 60
Tay Bridge, 57
The Pentagon, 175
USAF warehouses, 197
Walkways at the Hyatt Regency Hotel, 21
West Gate Bridge, 133
Willow Island Cooling Tower, 143
World Trade Center 1&2, 168
World Trade Centre 7, 180
Ynys y Gwas Bridge, 94

About the author:

Er. Dr. Niall MacAlevey is currently an independent consultant specializing in the analysis and design of reinforced and prestressed concrete structures, forensic engineering and the strengthening of concrete structures.

He is the founder of the firm "Shamrock Consultants", is a member of the Institution of Engineers of Ireland and is a registered Professional Engineer in Singapore. He graduated from University College Dublin, Ireland in 1987, and completed his M.Sc. degree in "Concrete Structures" at Imperial College, London. He completed his Ph.D degree at the Nanyang Technological University in 1997 on "The Strengthening of Concrete Structures" and later joined the academic staff there. He obtained a PGDipTHE (Post-Graduate Diploma in Teaching in Higher Education) from the National Institute of Education in 2001. He has worked for a number of consulting engineering firms and specialist prestressing subcontractors in London, Cambridge, Hong Kong and Singapore.

He can be contacted at niall@starhub.net.sg